D1453978

How to Confront Climate Denial

Literacy, Social Studies, and
Climate Change

James S. Damico and Mark C. Baildon

TEACHERS COLLEGE PRESS

TEACHERS COLLEGE | COLUMBIA UNIVERSITY
NEW YORK AND LONDON

Published by Teachers College Press,® 1234 Amsterdam Avenue, New York, NY 10027

Library of Congress Cataloging-in-Publication Data

Names: Damico, James S., author. | Baildon, Mark, author.
Title: How to confront climate denial : literacy, social studies, and
   climate change / James S. Damico and Mark C. Baildon.
Description: New York, NY : Teachers College Press, [2022] | Series:
   Research and practice in social studies | Includes bibliographical
   references and index.
Identifiers: LCCN 2022017573 (print) | LCCN 2022017574 (ebook) |
   ISBN 9780807767207 (paperback) | ISBN 9780807767214 (hardcover) |
   ISBN 9780807781159 (ebook)
Subjects: LCSH: Communication in climatology. | Climatic changes. |
   Climate change mitigation. | Climatic changes—Social aspects. |
   Curriculum planning.
Classification: LCC QC902.9 .D36 2022 (print) | LCC QC902.9 (ebook) |
   DDC 363.738/74—dc23/eng/20220622
LC record available at https://lccn.loc.gov/2022017573
LC ebook record available at https://lccn.loc.gov/2022017574

ISBN 978-0-8077-6720-7 (paper)
ISBN 978-0-8077-6721-4 (hardcover)
ISBN 978-0-8077-8115-9 (ebook)

Printed on acid-free paper
Manufactured in the United States of America

We dedicate this book to our children: Daniel Damico, Brady Baildon, and Taylor Baildon. May they continue to bring their many gifts to the world.

Contents

Preface

For the past 2 decades the two of us have explored challenges and possibilities that educators and students face when reading and evaluating online information sources as they investigate complex topics such as globalization, September 11 conspiracy theories, environmental racism, health-care disparities, and climate change. We have brought distinct yet overlapping interests and experiences to this effort. James is a former elementary and middle school teacher and a teacher educator and professor in literacy, culture, and language education; Mark is a former middle and high school social studies teacher in schools around the world (United States, Israel, Singapore, Saudi Arabia, Taiwan), a teacher educator and a professor in humanities and social studies education in Singapore, and is currently a professor in foundations of education in the United Arab Emirates.

Our decision to prioritize complex topics and social problems was shaped by James Martin's book, *The Meaning of the 21st Century* (2007), which outlines a host of interrelated "mega-problems" that threaten the planet. These intractable, urgent problems include global warming, excessive population growth, water shortages, destruction of life in the oceans, mass famine, the spread of deserts, pandemics, extreme poverty, growth of shanty cities, unstoppable global migrations, nonstate actors with extreme weapons, violent extremism, runaway computer intelligence, and cataclysmic war. What remains most striking about Martin's list is the likelihood that each of these problems or sets of challenges intensifies if global warming continues unabated.

We began teaching about climate change in 2008 (with the U.S. presidential race between Barack Obama and John McCain in full swing) and soon thereafter were inspired by Desmond Tutu, archbishop emeritus of Cape Town, South Africa, who identified the "tepid response of school-age and higher education to the global warming challenge" (Tutu, 2010, p. xvi). To confront "The Fatal Complacency," Tutu challenged educators to understand climate change as a global social-justice issue and to best use our hearts and heads to "guide what we do in the name of the human family" (p. xvi). Since that time we have interpreted Desmond Tutu's words as a summons to more deeply explore and articulate ways that global warming and climate justice might become more central to the work of educators.

Along with our colleague Alexandra Panos, we have considered ways university students evaluate the reliability of competing websites about climate change (Damico & Panos, 2016, 2018), including what happens when students dialogue across their divergent climate beliefs as they jointly evaluate climate-related webpages (Damico, Baildon, & Panos, 2018; Damico, Panos, & Baildon, 2018). We have also outlined ways to read a climate denial text (Baildon & Damico, 2011b), explored curricular and teaching possibilities in a "post-truth" era and in an "age of limits" (Baildon & Damico, 2019), and considered our own journey toward "ecojustice literacies" (Damico, 2021). All of these efforts have been grounded in an understanding of climate change as a socioscientific, rather than solely a scientific, topic. Climate change cuts across political, civic, geographic, economic, social, cultural, psychological, and historical dimensions (Damico et al., 2020; Klein, 2014), and across traditions in natural science, environmentalism, human rights, economics, politics, and religion (O'Brien, 2017) that require input from a range of scientific disciplines as well as socio-ecological systems (Newberry & Trujillo, 2019).

Over time, we have come to understand that our own teaching and learning about climate change must deal much more directly with *climate denial*: the rejection of the well-established scientific consensus about the causes and consequences of global warming *and* the rejection of the timely actions necessary to address these consequences and preserve a more stable planet for future generations. Given the likely devastating global consequences of climate change if not dealt with, we have come to identify climate denial as arguably the most consequential topic of our time.

This book is an attempt to chart a course for making climate denial a curricular and instructional priority in schools and other sites of learning. While we emphasize the academic content areas of social studies and literacy/ English language arts, climate denial is ripe for exploration and examination across other school subjects (science, mathematics, art, health, music, library/ media) in elementary through secondary school classrooms and across programs, departments, and schools in university settings. We hope a wide range of educators—teachers, parents, professors, administrators, and counselors, among others—who want to do more to address the climate crisis—find a way forward with this book. There is so much crucial and consequential work for all of us to do.

Acknowledgments

We owe a debt of gratitude to many people who helped shape this project. We begin with our colleague Alexandra (Alex) Panos. Alex began working with James in 2014 and has made many crucial contributions to the research and thinking that led to this book. Alex also played a key role helping develop the *Climate Denial Inquiry Model* in Chapter 2 and took the lead creating the visual of the model. We owe special thanks to Alex. We also received timely, useful feedback along the way from colleagues and friends. We thank Richard Beach, Brady Baildon, Deborah Chua, Gabrièle Abowd Damico, Rob Kunzman, and Michael Shafer for engaging thoughtfully with our ideas and offering clarifying questions and comments. We extend a special thanks to Loren Lybarger for his ongoing support of this project and for his many suggestions that nudged us toward greater clarity and comprehensiveness of our ideas. We also remain grateful to numerous thought and action leaders related to climate, social studies, and literacy we relied on to develop the ideas in this book. Their names and references to their work appear in the pages to follow. There is also an appendix to highlight resources that we consider essential to understanding and teaching about climate denial.

We thank Wayne Journell, series editor for the Research and Practice in Social Studies Series at Teachers College Press, for his enthusiasm and commitment to this project. We also thank Rachel Banks, associate editor, and Brian Ellerbeck, executive acquisitions editor at Teachers College Press, for guiding us with steady hands. Institutional support is also vital for a project like this and we extend our gratitude to the Indiana University School of Education–Bloomington for providing funding in the early stages of this project.

Our families sustained us throughout this effort and deserve special thanks. For James, Gabrièle Abowd Damico offered ongoing encouragement and wise counsel. She also continues to model what being grounded in life can look like, whether it is growing nourishing food in the community garden, tending to the many plants and flowers encircling our home, or nurturing relationships with all people in our lives. Daniel Damico offered spirited support all along the way. His reverence for nature and compassion for all plant and animal life is matched with joy-filled humor and affection for friends, family, and neighbors. Daniel also shares the lead vocal with Allen Davis on "Meet Us in the Street," a song written by James (Amigo Fields, 2021). The song is a conversation between adults who have had "our heads in the sand too long" and youth who have been leading the way to address climate change. A video accompanying the song can be found at http://y2u.be/TytLmjgOMuU.

For Mark, Rindi Baildon has been a source of inspiration and wisdom for her love of the outdoors and her passion for teaching elementary-age students to care for nature. His son, Brady, and daughter, Taylor, prod their father to continually see the world in new ways, to see that new possibilities lie ahead if we are open to them. We also acknowledge Carol Damico and Thomas V. Abowd and Anne Marie Abowd and Carol Baildon along with extended family members and friends. This list includes Brendan Shane, Nancy Markoe, and David Keller, who each played a pivotal role in shaping James's understandings and commitments to environmental action, gender equity, and civic engagement.

Finally, we are grateful that the two of us have cultivated and sustained an enriching collaboration for more than 2 decades. We have learned a great deal with and from each other that has been grounded in and enriched by all those who have crossed paths with us in classrooms, workshops, and conferences across a range of schools and communities. We consider ourselves fortunate to be educators.

We also acknowledge the publishers who granted us permission to draw from material that initially appeared in the following publications:

Baildon, M., & Damico, J. S. (2019). Education in an age of limits. *Journal of Curriculum Theorizing, 34*(3), 25–40. https://journal.jctonline.org/index.php/jct/article/view/823

Damico, J. S., Baildon, M., & Panos, A. (2018). Media literacy and climate change in a post-truth society. *Journal of Media Literacy Education, 10*(2), 11–32. https://digitalcommons.uri.edu/jmle/vol10/iss2/2/

Damico, J. S., & Panos, A. (2016). Reading for reliability: Preservice teachers evaluate web sources about climate change. *Journal of Adolescent & Adult Literacy, 60*(3), 275–285.

Damico, J.S. & Panos, A. (2018). Civic literacy as 21st century source work: Future social studies teachers examine web sources about climate change. Reprinted from *The Journal of Social Studies Research. 42*(4), 345–359. Copyright (2018), with permission from Elsevier.

Damico, J. S., Panos, A., & Baildon, M. (2018). "I'm not in the truth business": The politics of climate change with pre-service teachers. *English Teaching: Practice & Critique, 17*(2), 72–89.

Damico, J. S., Baildon, M., & Panos, A. (2020). Climate justice literacy: Stories-we-live-by, ecolinguistics, and classroom practice. *Journal of Adolescent & Adult Literacy, 63*(6), 683–691. https://doi.org/10.1002/jaal.1051

Introduction

How do we, as human beings, understand our relationships to each other and to all planetary species? What should be the goals of a society? Whom do we trust to navigate complex topics and issues? How should we act to address significant problems? How do we get where we want or need to go?

These questions are fundamentally about literacy and social studies. A literacy perspective highlights the meaning-making effort needed to engage with all different kinds of texts and information sources to help answer questions. Reading, writing, viewing, speaking, listening, and movement, including drama, are core literacy processes that facilitate this meaning-making. In social studies, the coin of the realm is content and that comes with a primary concern to develop, frame, and guide investigations into a potentially wide and diverse range of topics. As an inherently multidisciplinary field, covering the academic disciplines of history, geography, economics, psychology, sociology, and political science or civics, the social studies can be understood as offering a range of content maps to help orient the literacy practices needed to navigate different landscapes of significant topics and issues, including climate science denial and denial of the need for bold, transformative action to address climate change.

Narratives, or stories, are also core to literacy and social studies. Literacy educators dedicate a great deal of time to guiding students in reading, responding to, creating, and performing stories and in working with fictional accounts and real-life narratives. For social studies educators, particularly in history, narrative is a primary tool to help understand the past and its relationship to the present—ideally this includes teaching about how narratives are created, legitimated, and disseminated to serve different social and political purposes (Segall, 2006).

Climate denial is rooted in particular political, social, economic, and cultural myths, narratives, or "stories-we-live-by" (Stibbe, 2021). In contrast to a conventional understanding of "story" in which readers or viewers are more aware of the structure and content of what is being told, we are exposed to stories-we-live-by

> without consciously selecting them or necessarily being aware that they are just stories. They appear between the lines of the texts which surround us in everyday life: in news reports, advertisements, conversations with friends, the weather

forecast, instruction manuals or textbooks. They appear in educational, political, professional, medical, legal and other institutional contexts without announcing themselves as stories. (Stibbe, 2021, p. 5)

Several dominant *stories-we-live-by* include humans are the center of existence and separate from nature; the goal of a society is perpetual economic growth without limits; consumerism is the primary pathway to fulfillment; and nature is solely a resource to be used for human purposes (Stibbe, 2021). "Rugged individualism" is another prevailing myth with a strong foothold in the United States, one that aligns with stories about individual freedom, the inherent value of competition taking precedence over the common good, and the zero-sum story in which a person or group can only benefit when another person or group is disadvantaged or loses. One consequence is that a relatively small number of wealthy individuals can reap huge profits at others' expense, which produces greater inequality, sharper divides based on identity (especially racial identity), and harsher realities for more people in the United States and around the world (McGhee, 2021; Guterres, 2020).

The "we" in *stories-we-live-by* is not to suggest that all people in a society unwittingly acquiesce to these stories in the same way. Some individuals, cultural groups, and communities resist or defy these stories, and some have been doing it for decades or centuries. *Stories-we-live-by* are not universal, coursing through all cultures in the same way. Yet, *stories-we-live-by* are steeped in and shaped by social, political, and economic systems with corresponding values, principles, and commitments, which inevitably get "embedded deeply in the minds of individuals across a society" in such a way that they are not "immediately recognisable as stories" (Stibbe, 2021, p. 5).

This book takes aim at two kinds of *stories-we-live-by*. The first kind refers to systems-wide political, social, economic, and cultural stories that, if allowed to persist, will likely lead to catastrophic species and habitat loss across the planet. If governments, corporations, other institutions and groups, along with individuals, continue to uphold stories like *nature is a resource to be used and exploited* and *the primary goal of society is economic growth without limits,* then, for example, the devastating ecological and health impacts of deforestation in the Amazon can continue and are more easily justified. This is why *stories-we-live-by* "need to be exposed, subjected to critical analysis, and resisted if they are implicated in injustice and environmental destruction" (Stibbe, 2021, p. 5).

The second kind of story is about reliability and trust, how people determine what texts and information about topics of public significance they can trust. Having reliable information is crucial to reach informed conclusions and make sound decisions about any issues of consequence. Given the near endless streams of misinformation, scams, hoaxes, and unvetted, fake, or doctored information that circulate in online spaces, discerning who and what to trust must be a priority in literacy and social studies. Discerning the reliability of texts is a core literacy practice. In social studies, reliability or trustworthiness of information has increasingly become a key concept or topic in an increasingly hyper-partisan United States. Reliability and trust are core to collective and inclusive participation in a democracy. Whom we trust to provide information about important public issues like climate change determines how we think about those issues and our ability to effectively participate in public decision-making. Citizens in a democracy are empowered to pursue their individual and collective interests through argument, advocacy, deliberation, voting, and other forms of civic action, which rest upon stories of what constitutes reliable information and who should or shouldn't be trusted (Warren, 2018). This is why questions about reliability and trust are also fundamentally about subject matter content, how content is framed and investigated (as Chapter 3 highlights).

Reliability stories are also spread through society with people not necessarily being aware these are operating as stories. These *reliability stories-we-live-by* include (1) needing to always have the "other side" represented, (2) wanting more information or evidence to support an argument, and (3) reasoning primarily through one's own identity and perspectives (Damico, Baildon, & Panos, 2018). In contrast to a story like *humans are the center of existence and separate from nature,* these three reliability stories do not seem to justify or promote ecological harm. Yet, as we demonstrate in this book, they can contribute to, reinforce, or even amplify climate denial, so reliability stories also need to be analyzed critically and challenged.

PURPOSE AND AUDIENCE

Climate change and climate denial encompass political, civic, geographic, economic, social, cultural, psychological, and historical dimensions (Damico et al., 2020; Klein, 2014). In the context of schools, and in education more generally, climate change is an inherently transdisciplinary topic, providing opportunities for educators across academic disciplines and subject matter areas to make it a curricular priority. Given the backgrounds and teaching experiences the two of us bring to this project, we focus specifically on literacy/language arts and social studies where climate change has remained largely off the radar in both social studies (Kissling & Bell, 2019) and literacy education (Panos & Damico, 2021). This needs to change and this book offers one way forward. With a *Climate Denial Inquiry Model* (CDIM), as shown in Chapter 2, and classroom-based examples as a springboard to explore possibilities, we demonstrate how teachers and students can begin to confront climate denial as a multifaceted social problem. While we draw primarily on our work with undergraduate students, the CDIM was designed with middle and secondary level students in mind and can even be used with younger learners.

The CDIM highlights how teachers and students can use critical literacies alongside eco-civic practices of deliberation, reflexivity, and counternarration to discern underlying corporate, financial, and politically motivated roots of climate denial. Students will better understand efforts to misinform the American public, sow doubt and distrust of basic scientific knowledge, and erode support for evidence-based policymaking and collective civic action, so they can fully address the threat of climate change. These critical literacies and eco-civic practices can also help all of us examine the ways climate denial operates across different texts and in our own lives.

The CDIM supports student inquiries into the contexts of denial (economic, political, cultural-historical, sociopsychological, geographical, and media contexts) with an emphasis on helping students discern the reliability or trustworthiness of texts. In this way, the CDIM points to new reliability stories that can help students make determinations of reliability that are aligned more with notions of social trust grounded in evidence, deliberation, and counternarration. Ultimately, the two of us want to continue working and learning with and alongside educators and students to skillfully identify denial tactics, techniques, and texts and to critically examine how our individual and collective beliefs, values, experiences, and perspectives influence how we understand and engage with climate denial.

The Intergovernmental Panel on Climate Change (IPCC), the leading global authority on climate science, has provided a scope and timeline to prevent even more devastating effects of global warming. To limit warming to 1.5 degrees Celsius, global greenhouse gas emissions need to be roughly halved by 2030 and the world needs to achieve net-zero emissions by 2050 (the point when the amount of carbon dioxide removed from the atmosphere equals the amount that is produced). With this time frame, fully reckoning with climate denial in education entails developing a clear understanding of what climate denial is, what contexts cultivate and sustain it, and what tools and resources can be mobilized to confront it. The goal is to challenge the forms and facets of denial that impede the acceptance of climate science or thwart the large-scale and time-sensitive actions needed to reduce and "draw down" greenhouse gas emissions, primarily carbon dioxide and methane (Hawken, 2017). While forces of denial remain powerful in the United States and in other parts of the world, there are clear signs of resistance to climate denial as people and groups take action to demand a rapid and an equitable, just transition away from carbon-intensive fuels to address the climate threat. Part of this work, as the culminating chapter explores, is cultivating and enacting *ecojustice stories-To-live-by*.

Ultimately, to engage with climate denial, and the ecological and cultural crisis more broadly, we need to examine questions like, What stories are we living by? What stories are worth preserving? Which do we want to cast aside? And, what do we want our stories to be? Grappling with questions about how people can live together is essential to "the political classroom" (Hess & McAvoy, 2017), and exploring what we need to conserve and what we need to transform reflects "a pedagogy of responsibility" (Martusewicz et al., 2021). And to do this we need literacy tools, such as robust ways to read and analyze, along with social studies resources, including inquiry frameworks, key disciplinary perspectives, and clear civic education commitments. And working to identify and confront denial needs to take place in schools and classrooms, which "have emerged as a battleground in the American political war over climate change" as millions of children, youth, and young adults in the United States receive mixed, misleading, or false messages about climate science and its consequences (Worth, 2021).

A NOTE ABOUT TERMS

Climate Change and Global Warming

To signal how human impact has led to global temperature rise, we use the terms *climate change* and *global warming* interchangeably. We recognize that Frank Luntz, an advisor to President George W. Bush's administration, successfully persuaded the Republican Party to replace the term global warming with the "less frightening" term climate change. We are also aware that other terms like *climate disruption* and *climate destabilization* or more descriptive and arresting terms such as *climate collapse, climate chaos,* or even *climate catastrophe* are being used to describe climate-related events. We side with climate scientist Michael Mann (2021) that *climate crisis* is an apt term to describe the current challenges and to identify that difficult and significant decisions need to be made to address these challenges. As a result, we use *climate crisis* to name the broader social and political conditions brought on by global temperature rise. We also primarily use the term *climate change* to align our language with the way climate-related standards and curricula have been making their way into primary and secondary level schools in the United States and in other contexts.

As the following pages make clear, we find the term *anthropogenic*, or human-caused, climate change a bit misleading. It seems more accurate to understand and discuss climate change as industry-caused, industry-induced or, using a softer term, industry-led. The world is experiencing climate change primarily because leading coal, oil, and gas companies have denied climate science and continued to escalate greenhouse gas emissions despite having known for decades that this would very likely lead to current global warming conditions. The fossil fuel industry, of course, did not act on its own, as other industries, such as public relations companies and media corporations enabled their climate denial efforts (Brulle & Werthman, 2021). Law firms working for fossil fuel companies have done the same (Anang, 2021). Banks have also supported these companies with extensive fossil fuel financing, including several trillion dollars in the past few years (Kirsch et al., 2021). Insurance companies have also played a part with these denial efforts (Sherrington, 2021). The agricultural industry—namely, the largest meat and dairy corporations—have not only contributed significant greenhouse gas emissions themselves, but have also worked in tandem with the fossil fuel industry for several decades to undermine climate policy and stymie needed environmental regulations, primarily through the efforts of their lobbying organizations, such as the American Farm Bureau Federation (Banerjee et al., 2018). So, while an industry-caused climate change frame does not absolve individuals from our climate-related choices, especially those of us from major industrialized and carbon-intensive countries, the starting point for this book's examination of climate denial is at the industry level. Yet, this inevitably includes the relatively small number of extremely wealthy Americans who have been the primary funders of industry-led climate denial (Mayer, 2016).

Denial and Denialism

Kahn-Harris (2018) makes a distinction between denial and denialism. While denial includes an assortment of "humdrum deceptions and self-deceptions," denialism is "an expansion, an intensification, of denial" that represents "the transformation of the everyday practice of denial into a new way of seeing the world and . . . a collective accomplishment" (pp. 2–3). For Kahn-Harris, denial "hides from the truth" while denialism "builds a new and better truth" (p. 3). We appreciate Kahn-Harris's distinction, yet we use the term *denial* in this book with the cultural, collective, and sociological starting point that denial is "socially organized" (Norgaard, 2011; Zerubavel, 2002). Denial is also a core component of our particular "post-truth" moment characterized by a divisive politics in which appeals to emotion and ideology have been more influential than facts or expertise in shaping public opinion, making climate denial not only a rejection of scientific facts, but a refutation of the real-world or political implications of those facts (Fischer, 2019; Klein, 2011). To indicate that climate denial is actively promoted by groups and individuals, Worth (2021) uses the term "'climate denier' to describe groups or people who know (or who should know, based on their position or declared authority) that the conclusions of modern climate science are legitimate, but who nonetheless promote the idea that they are not."

Stories-We-Live-By, Frames, and Discourses

Stibbe (2021) outlines ten different language-based forms that stories take: ideology (how the world is or should be), framing, metaphor, evaluation (whether a story is good or bad), identity (what it means to be a particular kind of person), conviction (whether a particular description is true or false), erasure (what is not important), salience (what is important), and narrative (sequence of events; p. 17). While we do not emphasize all these particular forms in this book, the concept of *stories-we-live-by* overlaps with two related terms we do draw upon: *frames*, or cognitive structures that shape how we see the world (Lakoff, 2014); and *discourses*, or standardized ways that groups use language, images, and other forms of representation (Stibbe, 2021).

Literacy

We use the term *literacy* as an umbrella term to include school-based subjects that go by different names in kindergarten through high school classrooms (language arts, reading/writing, literature, English, composition, and media studies) and to describe meaning-making practices in particular subjects or academic disciplines—for instance, what it means to read, write, and communicate like a scientist or historian. At the university level, literacy also relates to coursework in schools of education (e.g., methods of teaching reading, writing, literature) as well as kindred types of courses across campus.

Text and Source

As we explore different examples of climate denial, we use the terms *text* and *source* interchangeably. Each refers to the particular unit we analyze, whether it is a website, YouTube video, social media post, or other source. The decision to use both terms reflects the ways our respective fields in literacy and social studies tend to use them ("texts" in literacy and "sources" in social studies).

Pronoun Use of *We*

We use the pronoun *we* to signal a collective sense of shared struggle involved with identifying and confronting climate denial. At times, the *we* indexes readers and citizens across the United States and beyond with the core arguments applicable to residents in other countries and contexts, especially people living in industrialized nations with high carbon emissions and some level of climate denial. Other times, the *we* refers to educators, or specifically to social studies and literacy educators. The *we* can also allude solely to the two of us, as in we hope these different uses of *we* are clear throughout the book.

OUTLINE OF CHAPTERS

Chapter 1, Climate Change, Denial, and Justice, provides the roadmap for the book as we answer questions about the climate science consensus, the causes and implications of climate change, how people in the United States understand climate change, why people deny the science consensus, what climate change education looks like in the United States, and how climate denial is tied to longstanding inequalities and injustices. The chapter concludes with a call for a robust, user-friendly model to help educators confront climate denial.

Chapter 2, Climate Denial Inquiry Model, heeds this call as we outline the core components of an inquiry model designed to help educators do this essential work. The chapter begins with examples of denial texts and we describe how these texts link to particular stories-we-live-by (Stibbe, 2021) that are shaped by and situated in larger social and cultural contexts of denial in the United States. These "swirling" denial contexts mix with core ecological philosophies, or "ecosophies" (Naess, 1995) of readers (educators and students) who can take up critical literacies and eco-civic practices of deliberation, reflexivity, and counternarration to better understand how mechanisms and techniques of denial work in and across texts. We then briefly describe how engaging in this work helps produce more justice-centered *reliability stories-To-live-by* as well as *ecojustice stories-To-live-by*.

Chapter 3, Web of Climate Denial, begins with the work of Dunlap and McCright (2011), who outline key components of a "climate change denial machine," a network of fossil fuel industry–led organizations, associations, and foundations that have challenged climate science and impeded policy action. We then explore how economic, political, cultural-historical, socio-psychological, media, and geographic contexts of climate denial coalesce with destructive stories-we-live-by and undermine climate science and necessary climate action. In many ways, this chapter offers essential social studies content for understanding climate denial and background knowledge for inquiry-based investigations. Chapter 7, in particular, builds directly from this foundation to chart potential inquiry pathways with these denial contexts in mind. Chapter 3 also sets the stage for inquiry explorations in Chapters 4 and 5 as each of these chapters explores the inquiry question, *Which sources about global warming do we trust and why?*—a question deeply rooted in denial contexts described in Chapter 3.

Chapter 4, Denial Texts and Critical Literacies+, explores how critical literacies can be mobilized to analyze denial texts. We ground this chapter with a four-phase instructional model to explore the inquiry question, *Which sources about global warming do we trust and why?* We examine an influential *climate science denial* text, highlighting how critical literacies can reveal core science denial techniques at play with this text. We also explore the relationship between this denial text and the larger web of denial contexts from Chapter 3. Next, we investigate several *climate action denial* texts. Here, critical literacies related to "discourses of delay" (Lamb et al., 2020) and "rhetorical frames" (Supran & Oreskes, 2021) are mobilized. We highlight the ways readers can analyze how language and beliefs in texts justify climate inaction or inadequate responses (discourses of delay) as well as how language is used to evoke (frame) particular ideas or make some perceptions of climate change more salient than others.

Chapter 5, Eco-Civic Practices of Deliberation and Reflexivity, explores the eco-civic practices of deliberation and reflexivity as tools students can mobilize to help them respond to climate denial texts. Teachers can design opportunities for students to collaborate and jointly investigate texts across their differences to bolster their knowledge about climate denial. This includes ways that teachers can intentionally guide students to be reflexive (or critically self-reflective) in their deliberations with any type of climate denial text.

Chapter 6, Confronting Denial Through Counternarration and Reliability Stories-*To*-Live-By, further develops the concept of reliability. After highlighting a model we developed to help secondary-level students more skillfully evaluate websites (Damico & Baildon, 2015), we highlight how the work of undergraduate students points to particular *reliability stories* the students draw upon as they evaluate climate change websites. Then, we mobilize the tools of counternarration to craft a different set of *reliability stories-To-live-by* to more directly confront climate denial. This includes sketching out initial ideas for a *Public Trust for Information.*

Chapter 7, Deeper Dive Into Climate Denial: Classroom Inquiries, builds from Chapter 6 to consider additional ways literacy and social studies educators can make climate denial more central to their work. Chapter 7 also builds directly from the "web of denial" contexts in Chapter 3 to consider a number of curricular possibilities or "inquiry pathways" framed as potential inquiry questions and topics to guide students into deeper investigations of climate denial.

Chapter 8, *Ecojustice stories-To-live-by*, charts how educators can transition from ecologically and socially destructive stories-we-live-by to three *ecojustice stories-To-live-by*. While by no means exhaustive, these ecojustice stories can guide curriculum, teaching, and learning in a range of settings.

Since we began teaching about climate change more than a decade ago, there has been a sizable shift in ecological awareness, understanding, and activism. Across the globe millions of people have taken to the streets, agitating for swift, bold, comprehensive policies, programs, and actions to address the climate emergency. Youth, in many ways, through organizations like the Sunrise Movement, Fridays for Future, and Extinction Rebellion, have been leading these efforts by challenging what they identify as the intersecting ecological and social injustices (fueled by our carbon-based economy that is based on ideas of limitless growth) and by pushing for transformative social and economic changes. Ultimately, as educators committed to confronting climate denial, we not only need to understand how denial works, we also need to offer learning pathways that lead to inquiry-based explorations of more just and sustainable futures, what moving past or through denial can look like.

Finally, while we have been working on climate change in social studies and literacy education for quite some time, we know there is much more the two of us need to learn and do in and out of the classroom. We invite educators—teachers, administrators, parents, community leaders, among a range of other stakeholders—to join us on this journey.

CHAPTER 1

Climate Change, Denial, and Justice

In this chapter we pose and respond to core questions about climate change, climate denial, and climate change education in science, literacy, and social studies. We also make the case that climate denial perpetuates inequality, as poor, marginalized, and vulnerable communities, who bear the least responsibility for global warming, suffer the most from both climate change and its denial. Confronting climate denial is essential to advance climate justice.

CLIMATE CHANGE

What Is the Scientific Consensus?

Western-based scientific inquiry pointing to anthropogenic, or human-caused, climate change began in the mid-19th century. As early as 1856, Eunice Newton Foote, an early climate feminist (Johnson & Wilkinson, 2020), through her own series of simple experiments with glass cylinders, thermometers, and an air pump, made the link between carbon dioxide and global warming. A few years later, in 1859, John Tyndall showed how water vapor, carbon dioxide, and methane block radiation. Climate research continued through the decades as different theories were tested. A more solid consensus about anthropogenic global warming (AGW) emerged in the 1980s. In 1988, NASA climate scientist James Hansen offered testimony to a U.S. Senate committee, stating that global warming was significant enough to conclude "with a high degree of confidence a cause-and-effect relationship to the greenhouse effect." He also testified that a warming planet was not just a distant threat. It was already happening and there was an urgent need for a comprehensive reduction of carbon emissions.

While Hansen's testimony in 1988 is often cited as an inaugural attempt to enlighten leading U.S. politicians, legislators, and the general public, robust engagement with this climate science began nearly a decade earlier, as Nathaniel Rich's account of the 1979–1989 period demonstrates. In *Losing Earth: A Recent History*, Rich (2019) explains how shortly after the first World Climate Conference in Geneva in 1979, the wealthiest nations of the world signed a statement to reduce carbon emissions. During the 1980s scientific consensus for AGW had been established and there was an emerging international political consensus to bring about a binding global treaty.

With Hansen's testimony in 1988, global warming and the greenhouse effect "emerged from academia and government offices to mingle with popular culture" (Schneider, 1989, p. ix) as *Time* magazine profiled climate scientists and the U.S. Congress was taking action to address the climate threat. With particular prescience, prominent climate scientist Stephen Schneider noted at the time, "Battle lines were being drawn for what promises to be one of the most important political debates of this—and the next—century: what can or should we do to avert the possibility of an unprecedented threat to the global environment, global warming?" (1989, p. ix).

Nineteen eighty-eight also marked the year when the World Meteorological Association established the Intergovernmental Panel on Climate Change (IPCC) as the leading scientific authority on global warming. Beginning in 1990, the IPCC has periodically published reports about current climate science understandings. Overall, findings and recommendations across the reports have been consistent. Greenhouse gas emissions are warming the planet at an unprecedented pace. In August 2021 the IPCC issued another major report updating its comprehensive review of climate science research. Their findings represent what the United Nations secretary-general called a "code red for humanity" and reinforced the notion that deep, immediate cuts in fossil fuel production are necessary to limit the impact of global warming.

Why Should People Trust the Scientific Consensus?

Trust is fundamental to what we know and believe about the world. Since we lack specialized knowledge on most complex matters (like climate change) and have little time to fact-check most information we encounter on a daily basis, we rely on others for much of what we know. We trust scientific or disciplinary knowledge because it is produced in a self-regulating community of professionals (experts) using rigorous methods to gather and analyze empirical evidence and critically vet knowledge claims (Fish, 2001; Oreskes, 2019). Knowledge-building in these communities is done collectively and critically to make sure errors or mistakes are corrected, claims are warranted (based on evidence), and agreed-upon methods and standards are used to ensure trustworthy knowledge is produced.

Heading into the third decade of the 21st century, there is a preponderance of scientific evidence that human-caused climate change is happening. This expansive knowledge base about global warming has been grounded in the natural sciences, with data and findings across scientific disciplines sometimes called Earth Systems Science (Mooney et al., 2013). The preponderance of evidence standard helps citizens, advocacy groups, researchers, policymakers, and others (e.g., educators and students) "express themselves explicitly and with reasonable precision regarding the degree of certainty or uncertainty that they themselves associate with a given scientific assertion or chain of evidence" (Weiss, 2003, p. 27). This is an especially important standard in adjudicating public "controversies" that involve scientific claims and evidence, as well as in their denial. As such, it offers a way to understand the scientific consensus about climate change as well as address the denial of climate science. This evidentiary base from Western-based science also aligns with non-Western, Indigenous knowledge and wisdom traditions, what Goonatilake (1998) has called "civilizational knowledge."

Put simply, the primary reason to trust climate science is because what we know about global warming is based on extensive, well-developed, rigorous social and institutional processes and procedures (Oreskes, 2019; Sinatra & Hofer, 2021). This includes knowledge and the ecological wisdom amassed across centuries through Indigenous traditions, practices, and models of sustainability. Placing our trust in these rigorous scientific traditions and safeguards is much different than placing trust in people, groups, or organizations whose livelihoods benefit, or depend on, not accepting the climate science consensus, such as those who benefit from ties to the fossil fuel industry.

What Are the Causes of Climate Change?

The far-reaching effects of the global warming trend have led scientists to contend that humanity's profound influence on the Earth's surface, oceans, and atmosphere has ushered in a new geological epoch called the Anthropocene. Scientists disagree about an official starting date for the Anthropocene, or tipping point of human influence over the environment, alternatively pinning the origins to the advent of agriculture thousands of years ago, the beginning of the Industrial Revolution in the early 1800s, or around 1950 to mark the "Great Acceleration," a time of rapid, global industrialization as world population along with carbon and methane pollution began to spike significantly.

Since the beginning of the 18th century, the annual global average temperature has increased more than 1 degree Celsius (1.9 degrees Fahrenheit). Before humans began to burn coal, gas, and oil, the atmosphere routinely contained about 280 parts per million (ppm) of carbon dioxide. This meant that for every million molecules in the atmosphere, 280 of them were carbon dioxide. The upper threshold to maintain climate equilibrium is considered to be 350 million ppm, yet the National Oceanic and Atmospheric Administration (NOAA) documented in May 2021 that the world was approaching 420 ppm, as fossil fuel use continued to expand throughout the world, with China, the United States, the European Union, India, and Russia the countries most responsible for skyrocketing atmospheric carbon (Friedrich et al., 2020).

Put simply, our current state of global warming is tied to industry-induced greenhouse gas emissions that stem directly from the continued burning of fossil fuels despite evidence of its destructive effects. This points to the social, cultural, and economic roots of the climate crisis in which particular metaphors, narratives, or stories justify continued political and economic programs and practices that produce ecological harm (Bowers, 2001; Klein, 2014; Kimmerer, 2013; Martusewicz et al., 2021; Stibbe, 2021).

What Are the Effects?

Our planet's climate is in the midst of radical, transformative change. The scientific data is clear, comprehensive, and unequivocal. As a result, the planet is in the midst of a long-term trend: An escalating global temperature is leading to rising sea levels; the loss of glaciers, permafrost, and sea ice; an appreciable spike in heat waves; the expansion of deserts; changes in seasonal patterns (growing seasons, bird migration, insect infestations); a higher frequency/intensity of droughts, flooding, and fires (Environmental Protection Agency, 2021); and animal species loss and mass extinction events. The effects of climate change pose significant threats to our social and political systems, challenging food and water supplies, health-care provision, transportation systems, public policymaking, and social service provision for growing vulnerable populations, such as climate refugees (Dryzek et al., 2011; Flavelle, 2021).

In terms of climate impacts on human life, scholars are also examining the "mortality cost of carbon" through calculations of how many future lives will be lost or saved depending on greenhouse gas emission reductions in the coming decades (Bressler, 2021). Moreover, the widespread and severe damaging impacts of global warming has led an independent, global panel of legal experts to formally designate *ecocide*—defined as "unlawful or wanton acts committed with knowledge that there is a substantial likelihood of severe and either widespread or long-term damage to the environment being caused by those acts"—an international crime (Ecocidelaw, 2021).

What Can Be Done?

Project Drawdown is a comprehensive resource for ideas about what can be done about global warming (https://www.drawdown.org/about). (The term *drawdown* signals the time when greenhouse gas levels in the atmosphere stop rising and begin to decline). Based on the work of many scholars, scientists, policymakers, business leaders, and activists, Project Drawdown provides a framework of solutions based on three interconnected areas of climate action: (1) reduce sources of emissions (to achieve net zero); (2) support carbon sinks to assist nature's carbon cycle; and (3) improve society through health and education to foster sustainability and equality for all. Their website (https://drawdown.org/) offers concrete solutions that cut across multiple sectors of society (e.g., electricity; food, agriculture, and land use; industry; transportation; land, coastal, ocean, and engineered sinks; health and education). Of note, these three interlocking sets of solutions already exist to stop catastrophic climate change "as quickly, safely, and equitably as possible" (n.d); in their view, there is no need to wait for new innovative, technological solutions to accomplish what needs to be done.

While there is debate about the best solutions to address global warming across particular contexts, what remains undeniable is the need to decarbonize society, to radically reduce greenhouse gas emissions and transform the global economy with much greater energy efficiency, with a rapid transition to a net-zero carbon global economy (again with net zero defined as a state when greenhouse gases entering the atmosphere are balanced by their removal from the atmosphere). More specifically, to limit global warming to 1.5 degrees Celsius, global greenhouse gas emissions need to be roughly halved by 2030 and the world needs to achieve net-zero emissions by 2050 (IPCC, 2018). To meet this ambitious goal, the phrase and associated movement called "Keep It in the Ground" provides a metric to clarify what needs to be done: World leaders need to immediately halt all new fossil fuel development, advance a just transition to renewable energy, and manage the decline of the fossil fuel industry (#KeepItInTheGround, n.d.). The International Energy Agency has made a similar case, reporting that countries need to stop the approval of new coal-fired plants and new oil and gas fields in order to avert the most severe climate change effects (IEA, 2021). (Of note, fossil fuel companies, among other corporations, have been more likely to embrace the 2050 goal rather than take aim at the 2030 target with its more near-term timeline. The simple math here is that commitments farther into the future attenuate the accountability for the substantive, widespread changes needed in the short term.)

CLIMATE DENIAL

For decades major oil companies have known that their continued extraction and burning of fossil fuels accelerates global warming. The science was not controversial or complicated. Carbon dioxide is a heat-trapping greenhouse gas and when the world emits more of it from burning oil and coal than nature can remove, the amount of atmospheric carbon dioxide rises and the global temperature rises in turn. Rather than lead a transition away from these high-polluting energy sources, the fossil fuel industry led a decades-long campaign to sow doubt and confusion about the basic science and the clear, ever-deepening consensus about human- or industry-caused climate change (Freese, 2020; Oreskes & Conway, 2010). Its own business model hummed along, maximizing profits and generating vast amounts of corporate wealth while ignoring the escalating social, cultural, and ecological harm caused by its continued practices. Journalist Kate Aronoff (2021) offers a succinct summary of these actions: "The world's most prolific polluters have spared no expense over the last several decades to obstruct climate action and stymie efforts at every level of government."

Industry-led climate denial tactics across the years have been remarkably successful in delaying the needed action to address the escalating climate crisis. These tactics include attacks on the integrity of particular climate scientists and attempts to discredit or foster distrust of any institution that has acknowledged the scientific consensus (major scientific organizations, universities, media, and the government). Industry-led climate denial efforts have also forged and cemented ideological alignments with corporate-friendly politicians and political parties that have preserved corporate interests and have fueled political polarization and tribalism across society (Freese, 2020). Fossil fuel corporations did not act on their own; the public relations industry, along with major media corporations, have been dependable accomplices in denial.

Denial or rejection of the climate science consensus takes different forms and moves across different phases. Michael Mann (2012), the renowned climate scientist, noted six stages of climate science denial as expressed in the following arguments:

1. CO2 is not actually increasing.
2. Even if it is, the increase has no impact on the climate since there is no convincing evidence of warming.
3. Even if there is warming, it is due to natural causes.
4. Even if the warming cannot be explained by natural causes, the human impact is small, and the impact of continued greenhouse gas emissions will be minor.
5. Even if the current and future projected human effects on Earth's climate are not negligible, the changes are generally going to be good for us.
6. Whether or not the changes are going to be good for us, humans are very adept at adapting to changes; besides, it's too late to do anything about it, and/or a technological fix is bound to come along when we really need it. (p. 123)

These stages align with a taxonomy of the climate misinformation landscape with five main categories: it's not real, it's not us, it's not bad, climate solutions won't work, and the experts are unreliable (Coan et al., 2021). With these stages we can see how outright science denial can evolve into a begrudging acceptance of the science yet continue to reject the idea that a response to the escalating crisis requires changes in government policy, industry practices, or even individual behaviors. Stanley Cohen (2001) offers another way to understand different forms or "states" of denial. While Cohen developed his three-part framework analyzing genocide, his ideas apply to climate denial and inaction. The first form is *literal* denial, the rejection that something is happening (or lying), and is best captured with Mann's first stage above. Cohen's second form is *interpretive* denial in which facts are distorted or "spun" into unwarranted interpretations, as represented in Mann's second through fifth stages. The third form for Cohen is *implicatory* denial, the refusal to acknowledge and act upon the personal, moral, and political implications of established facts. Seeds of this type of denial are evident in Mann's sixth stage; while there is acceptance of climate science, the imperative to act is rejected.

It also bears noting that outright climate science denial has become less pervasive than interpretive or implicatory denial. Over the past decade big oil companies, for example, have moved away from literal climate denial to emphasize the ways the industry is making progress in addressing climate change or reinforcing its case that the industry is integral to society (Brulle et al., 2019). Fossil fuel companies use "greenwashing" and "discourses of delay" to shift the onus of responsibility away from them and to slow down or stymie necessary action (Lamb et al., 2020) or to, again, remind us about the essential contributions that fossil fuels make to society (Brulle et al., 2019). Put another way, the language of adaptation, conservation, and innovation promotes inaction, and that this shift from outright science denial to intensified efforts to downplay, distract, deflect, divide, and delay represents a "new climate war" (Mann, 2021). This includes industry's efforts to embrace "non-solution solutions" like natural gas, clean coal, and geo-engineering, which do not necessitate transitioning from fossil fuels as soon as possible.

Writer Alex Steffen (2017) uses the term "predatory delay" to describe how fossil fuel companies have continued to make ecologically unsustainable decisions that worsen injustices in order to maximize their wealth. The term *sustainable* also merits scrutiny for potential greenwashing. Monbiot (2017) highlights how "sustainability" can morph into "sustainable development" then "sustainable growth" and then "sustained growth," which, in practice, can be antithetical to "sustainability." As one indicator of denial as delay, many U.S. House Republicans, amidst a series of extreme weather events in 2021, agreed that global warming was happening but contended that a continued reliance on oil, coal, and gas was needed to prevent economic harm (Friedman & Davenport, 2021). In Chapter 2, we parse climate denial into *science denial* (rejection of climate research consensus) and *action denial* (not taking necessary action to address the climate crisis). Finally, while concerns are warranted about the potential economic impact of a transition from fossil fuels to renewable energy sources (the fossil fuel sector employs many people, and many economic sectors are fossil fueled), any perspective or proposal that falls short of acknowledging and addressing the urgent need for a rapid transition from fossil fuel extractivism points to *climate action denial*.

How Do People in the United States View the Climate Consensus?

With a team of psychologists, geographers, political scientists, statisticians, pollsters, and communication scientists, the Yale Program on Climate Change

Communication has collected national survey data to track public opinion about climate change since 2008. Their initial study, "Global Warming's Six Americas" (Leiserowitz et al., 2009), found that Americans can be categorized into six distinct groups based on their beliefs and attitudes about climate change—Alarmed, Concerned, Cautious, Disengaged, Doubtful, and Dismissive.

The Alarmed segment is most convinced by the scientific evidence that human-caused climate change is happening and considers it an urgent threat requiring bold new climate policies. Those categorized as Concerned similarly think climate change is happening, yet view it as a more distant threat in time and space and as a lower-priority public policy issue. The Cautious, while aware of climate change, haven't made up their minds about the seriousness of the problem. The Disengaged know little about climate change, while the Doubtful segment of the American population thinks climate change may be part of natural climate cycles or do not deem it a serious risk. Dismissives think global warming is a hoax, are more prone to conspiracy theories that dismiss climate science, and are most active in denying the need for climate-based actions.

Goldberg et al. (2020) found these six categories still hold more than a decade later, with the following changes in the percentages that make up these six groups:

- Alarmed segment has almost tripled in size (11% to 31%).
- Concerned segment has remained fairly stable (30% to 26%).
- Cautious segment has decreased (24% to 16%).
- Disengaged segment has largely stayed the same (8% to 7%).
- Doubtful segment has decreased (15% to 10%).
- Dismissive segment has also decreased (12% to 10%).

Figure 1.1 depicts the 2020 percentages of the study.

The latest survey found the Alarmed segment to be at an all-time high (31%). Yet, while there has been an overall shift to more Americans embracing climate science and the need for more timely climate action, close to half of the population (43%) remain cautious, disengaged, doubtful, or dismissive of the science—leaving the United States still a long way from embracing the reality of climate science. Moreover, there is a sizable gap between accepting or "believing in" climate science and the willingness to take swift, substantial action to address the climate threat.

Yet, some argue that this is not necessarily cause for defeatist or doom-based perspectives about the possibility for impactful climate action. For example, given that roughly 75% of people are not "alarmed," one could conclude that there is not enough time to respond to the climate crisis, which feeds perspectives such as "it's futile to try" or "let's prepare for the worst." Yet, the Yale Climate Change Communication survey findings point out that the dismissives represent about just 10% of respondents. For climate change

Figure 1.1. Global Warming's Six Americas (Goldberg et al., 2020)

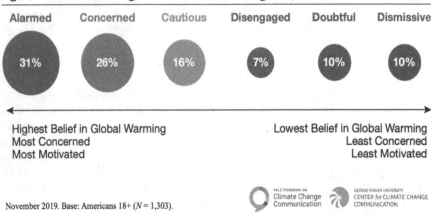

November 2019. Base: Americans 18+ (*N* = 1,303).

scientists and communicators like Katharine Hayhoe and Michael Mann, this means an overwhelming majority of people in the United States are NOT actively countering climate science and are not rigidly opposed ideologically to climate action and, thus, could be receptive to a range of climate-based policies and actions that point to improved social, economic, and health-based benefits tied to decarbonization.

Why Do People Deny Climate Change?

One common explanation for why people deny the reality and existence of climate change is that they lack the relevant information or knowledge. The common remedy to address this gap is to communicate climate science information through books, websites, podcasts, social media, research reports, and other sources. Though this "information deficit model" has received much criticism for not attending to the complex ways knowledge, beliefs, attitudes, and behaviors intermingle with a polarized topic like climate change, this model remains a key component of climate change communication practice and research (Suldovsky, 2017). This model, however, is more effective when issues are not entwined in politics, identity, and morality or when the science does not point to the need for urgent and far-reaching action (Hayhoe, 2021; Lakoff, 2014).

Sociologist Kari Norgaard (2011) demonstrates the limitations of the information-deficit model by describing the ways climate denial is constructed in a community. Her study of a rural Norwegian community during an unusually warm winter highlights how the community did not integrate their clear knowledge of climate change into their everyday lives. Working with the concept of the *social organization of denial* (Zerubavel, 1997, 2002), Norgaard shows how the people in this community grasped climate change impacts yet used a repertoire of strategies to detach these impacts and the need to act from their own social, political, and economic lives. The community members distanced themselves from responsibility "to assert rightness or goodness of actions, to maintain order and security, and to construct a sense of innocence in the face of the disturbing emotions associated with climate change" (Norgaard, 2011, pp. 11–12). Put another way, climate realities conflicted with the story of the community's or individual's "goodness," a story of the need to maintain order/security/status quo, and a story of innocence.

The social organization of denial, and climate denial more specifically, are also shaped by psychological processes as brains are "wired for denial" (Marshall, 2015). These processes include *identity-protective cognition*, a type of motivated reasoning in which the goal is to maintain one's status or membership in a group; *selective perception,* or biased assimilation, whereby people tend to interpret new information or evidence as consistent with their prior beliefs; *confirmation bias*, the inclination to search for information that confirms one's prior beliefs; and *disconfirmation bias*, the tendency to more closely scrutinize any information that counters one's prior beliefs (Druckman & McGrath, 2019).

In the context of a community, then, these psychological processes can further support stories a community might value—for example, about the need to maintain order, security, and the status quo, or even its innocence. The ways communities understand and respond to climate change are also influenced by dominant discourses in the larger society. Deep cultural assumptions (Bowers, 2001), which often take the form of "stories-we-live-by" (Stibbe, 2021), shape how individuals and groups interpret, understand, and respond to the climate crisis. To confront climate denial, there is a need to challenge stories—such as humans are autonomous individuals, the environment is a resource to be used (and exploited), and unending progress (liberal and economic) is the primary goal of society—and replace these ideas with a more ecologically grounded social justice vocabulary (Bowers, 2017) and more beneficial stories (Stibbe, 2021).

As upcoming chapters will explore, climate denial is financed and disseminated through a "denial machine" as fossil fuel companies and well-financed foundations and organizations disseminate misinformation and disinformation (Dunlap & McCright, 2011) and more destructive stories (Stibbe, 2021) through a corporation-friendly media (and social media) environment. For example, a recent report by Avaaz (2021), a nonprofit organization that promotes global activism, points to an evolving landscape of climate misinformation on social media, particularly on Facebook, where interactions around posts of climate misinformation out-perform factual climate posts in the *New York Times* and the *Washington Post*. In the report's executive summary, they note:

Tactics to prevent climate action are shifting away from outright climate change denialism, and toward more subtle misinformation narratives that promote "inactivism" . . . by asserting that it is too late to act on climate, promoting ineffective solutions to the climate crisis, or baselessly attacking effective forms of climate change mitigation. For instance, the most common misinformation narrative in our dataset did not deny the existence of climate change, but instead framed renewable energy sources like wind as being inefficient, unreliable, or otherwise harmful. (p. 1)

These findings fit into a taxonomy of the climate misinformation landscape expressed as five core arguments: It's not real, it's not us, it's not bad, climate solutions won't work, and the experts are unreliable (Coan et al., 2021). Ultimately, the short answer to the question, *Why do people in the United States deny climate change?* is due to decades-long fossil fuel industry efforts to deny climate science and delay climate action. In late 2021, these denial efforts showed no signs of stopping. In testimony with the U.S. House Committee on Oversight and Reform, executives from Chevron, ExxonMobil, BP, and the American Petroleum Institute (API) refused to acknowledge the existence and influence of their climate denial disinformation campaigns despite the clear, well-documented, and publicly known evidence that has established, in great detail, how their denial tactics were put into practice.

CLIMATE CHANGE EDUCATION IN THE UNITED STATES

The educational approaches to address climate change in the United States include school-based programs in kindergarten through high school and in university settings as well as a range of initiatives in less formal contexts, such as environmental centers and museums and through other community efforts (Henderson & Drewes, 2020). These efforts tend to emphasize the science of climate change (e.g., Duggan-Haas & Ross, 2017; Fester & Valenzuela, 2021; Shepardson et al., 2017; Zabel et al., 2017). Given that the educational system in the United States allocates control to each of the states, and local community school boards retain a great deal of authority to shape what happens in public school classrooms, it comes as no surprise that climate change education in schools varies considerably and remains a mixed bag (Henderson & Drewes, 2020).

What About Climate Education in Science Classrooms?

Climate change is included in the Next Generation Science Standards (NGSS), which were an attempt to bring some cohesion to science education across the United States. Yet, much control was left in the hands of state education systems and local school districts to determine how to address the standards (Chen, 2018). While some states and individual communities advance a more robust climate change agenda for schools, others advocate either avoiding or reshaping (or watering down) curricular expectations and priorities. New Jersey became the first state to incorporate standards for teaching climate change across multiple subjects in a coordinated and systematic way (Branch, 2020), and the state of Washington passed a 2018 law investing $4 million in science teacher training that included teaching climate science (Kagubare, 2019). States less willing to adopt the NGSS standards or prioritize climate change in the curriculum have included Texas, Ohio, Pennsylvania, Virginia, North Carolina, and Florida.

The slow or nonexistent adoption of the NGSS standards and teaching of climate change in some states is linked to the fact that some state economies and employment are more heavily reliant on the fossil fuel industry, and fossil fuel industry trade groups have been spending large amounts of money to influence curriculum in states like Texas and Ohio (Pilon, 2021; Worth, 2021). Across the United States, a well-funded and sustained attack on teaching climate change in schools includes efforts to block the NGSS standards in state legislatures and efforts at the local level to make climate science controversial, advance climate denial, and foster uncertainty about the science (Branch et al., 2016).

Seeing an opportunity to influence science education, the Heartland Institute, a politically conservative think tank and key cog in the "climate change denial machine" (Dunlap & McCright, 2011), sent 300,000 copies of the book *Why Scientists Disagree About Global Warming* (Idso et al., 2015) to K–12 and college science educators along with a DVD that challenges climate science. Key claims of the book include: There is no scientific consensus; the report from the Intergovernmental Panel on Climate Change (IPCC), the leading climate science authority in the world, is not credible and its report is based on flawed data, models, and projections; and the scientific community is using unreliable circumstantial evidence regarding the melting of polar ice caps, sea-level rise, and extreme weather events. The introduction to the book unequivocally states, "Probably the most widely repeated claim in the debate over global warming is that '97% of scientists agree' that climate change is man-made and dangerous. This claim is not only false, but its presence in the debate is an insult to science" (Idso et al., 2015, p. 1). We will further explore this claim in Chapter 2 and the role of the Heartland Institute in Chapter 4.

It also bears noting that fossil fuel industry influence in education is nothing new. For decades stretching back to the 1920s, the pro-corporate agenda of the fossil fuel industry has been disseminated in K–12 classrooms and in higher education (Westervelt, 2021; Worth, 2021). Investigative journalists Amy Westervelt and Dharna Noor, in a podcast series called The ABCs of Big Oil, chronicle the inroads oil companies have made in schools. Katie Worth, in the book *Miseducation: How Climate Change Is Taught in America* (2021), chronicles the ways fossil fuel corporations, state legislatures, school boards, conservative think tanks and lobbyists, and textbook publishers have been sowing uncertainty and distrust about climate science and shaping climate change education in the United States.

What About Climate Change in Social Studies and Literacy Education?

The National Council for the Social Studies (NCSS) has called for climate change to be included in the curriculum. Sponsored by the College and University Faculty Assembly, Resolution #19-02-2 states that the NCSS, which is "concerned with the education of our diverse, multicultural citizenry, should emphasize the cultivation of ecological citizenship" (National Council for the Social Studies, 2019).

The National Council of Teachers of English (NCTE) has also called for climate change to be included in English/language arts classrooms. Its *Resolution on Literacy Teaching on Climate Change* encourages teachers to resist the politicization of climate science by evaluating curricular texts for scientific credibility; lead students to engage thoughtfully with texts focusing on social and political debates surrounding climate change; and work with teachers in other fields to implement interdisciplinary instruction on climate change and sustainability (NCTE, 2019). The ELATE Commission on Climate Change and the Environment in English Education (c3e3) further advances this work (Mayo & Novack, n.d.).

Yet despite the scope and urgency of the climate crisis and widespread denial in the United States, there is little evidence to suggest that climate change or climate justice has become a priority in social studies or literacy education. Kissling and Bell (2019) highlight how environmental issues have been largely ignored in social studies standards, textbooks, classroom practice, and research, which have focused steadfastly on human lives and society in ways that marginalize and diminish the Earth and all that is not human (Bigelow, 2014; Houser, 2009; Kissling & Bell, 2019). Similarly, in social studies teacher education, there has been little emphasis on preparing social studies educators to teach about the Earth, sustainability, or climate change (Kissling et al., 2017). Key barriers that make it challenging for social studies educators to engage with ecological issues include views that this is the domain of science education; that social studies educators are ill prepared to teach this content; that curriculum demands are already bursting at the seams; and that environmental issues are controversial and challenging to teach (Kissling & Bell, 2019).

Environmental issues have also been largely absent in the fields of literacy and language arts education. Panos and Damico (2021) conducted a systematic review of journal publications and conference presentations from 2008 to 2019 across the three leading literacy-related professional organizations—the International Literacy Association (ILA), the National Council of Teachers of English (NCTE), and the Literacy Research Association (LRA)—to find that climate change, and environmental issues more broadly, have been largely off the radar for these organizations and their publications and conferences.

This is not to suggest that there have been no calls to prioritize climate change or environmental justice issues in social studies and literacy education and that there have not been models to consider. D. W. Orr's (1992) collection of essays, titled *Ecological Literacy: Education and the Transition to a Postmodern World*, as well as his book, *Earth in Mind: On Education, Environment, and the Human Prospect* (1994), along with two books by C. A. Bowers, *Educating for Eco-Justice and Community* (2001) and *Revitalizing the Commons: Cultural and Educational Sites of Resistance and Affirmation* (2006), have charted a way forward. Other work has done the same, including Houser (2009), who has advocated for environmental approaches in civic education. A book that advances critical pedagogy with an ecoliteracy perspective is Kahn's (2010) *Critical Pedagogy, Ecoliteracy, and Planetary Crisis: The Ecopedagogy Movement*. The book *EcoJustice Education: Toward Diverse, Democratic, and Sustainable Communities*, now in its third edition (Martusewicz et al., 2021), repurposes social education in an age of ecological crisis, making linkages between ecojustice and the role of culture, value hierarchies, racism, and sexism. A companion text with a teacher audience in mind is *Teaching for EcoJustice: Curriculum and Lessons for Secondary and College Classrooms* (Turner, 2015). Barton and Ho (2021) dedicate a chapter to environmental justice in their book *Curriculum for Justice and Harmony*. Others are also highlighting the ecological crisis in social studies education (Garrett, 2019) including in venues like the *Visions of Education* podcast (Krutka, 2020).

Another useful resource, *Teaching Climate Change to Adolescents: Reading, Writing, and Making a Difference* (Beach et al., 2017), explores four curricular areas—Indigenous and postcolonial perspectives, capitalism and consumerism, environmental literature, and human-based systems—that point to issues-based instructional possibilities in both English language arts and social education. Another key teacher-friendly text, *A People's Curriculum for the Earth: Teaching Climate Change and the Environmental Crisis* (Bigelow & Swinehart, 2014), includes instructional ideas (e.g., role plays, poems, graphics, simulations) that tackle a range of ecological justice issues (e.g., climate change, water, food, pollution, energy) in a multidisciplinary way with accessible examples for teachers. The Zinn Education Project companion website (https://www.zinnedproject.org/) and the "Climate Change and Justice" unit from the Choices Program at Brown University (https://www.choices.edu/curriculum-unit/climate-change-questions-justice/) also offer a range of instructional resources.

Despite these noteworthy examples, the problem of scarce climate change curricular resources in social studies and literacy education persists. Consider another influential curricular resource in social studies, the *College, Career, and Civic Life (C3) Framework for Social Studies State Standards* (NCSS, 2013). Although "human-environment interaction and place" are identified as core geographic concepts, and "an environmental perspective that views people as living in interdependent relationships within diverse environments" is encouraged to help students understand the world as "a set of complex ecosystems interacting at multiple scales that structure the spatial patterns and processes that influence our daily lives" (p. 40), there is little to support inquiry into the social, civic, political, economic, geographic, or historical contexts of climate change or climate denial.

CLIMATE DENIAL, INEQUALITY, AND JUSTICE

In previous work we have called for a climate justice literacy framework based on three core understandings: (1) climate change is a complex, socioscientific problem; (2) climate change is mediated by a complicated, diverse array of motivated digital texts and readers; and (3) climate change is about (in)justice (Damico et al., 2020). In this book we establish a clearer, tighter link between climate denial and climate justice. The research is unmistakable: The most devastating consequences of climate change, such as food and water scarcity, habitat loss, dislocation or forced migration, are unevenly felt around the world, with poorer countries and communities and women (Nagel, 2016) disproportionately bearing the brunt of climate effects (Levy & Patz, 2015). Greenpeace, a nongovernmental environmental organization, even uses the term "fossil fuel racism" to describe how the burning of fossil fuels has led to disproportionate negative impacts in Black, Brown, Indigenous, and poor communities.

To help grasp the global scope of climate-related inequality, Oxfam (a major nonprofit group of charitable organizations) and the Stockholm Environment Institute compiled a report with the staggering finding that across a 25-year period (1990–2015) the wealthiest 1% of the global population (those earning more than $100,000 USD a year) were accountable for *twice* as much carbon dioxide emissions as the poorer *half,* or 50% of the world's population (Oxfam International, 2015). The climate impact "gap" is also notable within the United States (Frosch et al, 2018), with communities of color and low-income communities more likely to be affected by climate change–related events like flooding (Chakraborty et al., 2019) and heat waves linked to urban heat islands (Jesdale et al., 2013). Put simply, climate change is a threat multiplier that exacerbates existing injustices and vulnerabilities (Johnson & Wilkinson, 2020), an understanding that resonates deeply with decades of environmental justice work (Bullard, 2020; Sze, 2020).

This is not to suggest, however, that only frontline communities in the United States are experiencing adverse climate impacts. As global air and ocean temperatures rise, heat waves, flooding, wildfires, and droughts increase in number and intensity, with more people seeing and experiencing climate-related impacts. Based on a number of climate change indicators documented by the Environmental Protection Agency (2021), there is no city, rural community, or small town in the United States not affected by the climate crisis. For example, repeated extreme weather events especially threaten low-income and small rural communities, causing businesses and residents to leave, further shrinking tax bases and making it harder to provide basic services that would help people face future climate shocks (Flavelle, 2021).

Climate denial can also be situated within a longer historical trajectory of colonial injustices in the United States—namely, the forced expulsion and slaughter of Indigenous people along with theft and exploitation of Indigenous land and resources. The project Gesturing Towards Decolonial Futures points to four denials linked to "an inherently violent colonial habit of being":

- denial of systemic violence and complicity in harm (the fact that our comforts, securities, and enjoyments are subsidized by expropriation and exploitation somewhere else)
- denial of the limits of the planet (the fact that the planet cannot sustain exponential growth and consumption)
- denial of entanglement (our insistence in seeing ourselves as separate from each other and the land, rather than "entangled" within a living wider metabolism that is bio-intelligent)
- denial of the depth and magnitude of the problems that we face (https://decolonialfutures.net/4denials)

Climate denial compounds climate-related injustices because it impugns or dismisses the need to address the causes and consequences of global warming. It is a refusal to recognize that the wealthiest economies and populations contribute the most to global climate change, while the consequences of climate change, in terms of the impacts of extreme weather events, environmental and health consequences, and the impact on basic human rights (e.g., rights to security, livelihood, standard of living adequate for health and well-being) disproportionately affect lower-income countries and poorer communities (Levy & Patz, 2015). Moreover, climate denial can lead the wealthiest people across the globe, the primary emitters of carbon by a staggering margin, to sidestep a moral imperative to reduce their carbon emissions.

Understanding climate denial through a justice lens is also important for another reason. Big oil companies are making social justice more central with their "discourses of delay" by either warning that a transition from fossil fuels will negatively impact poor and marginalized communities or through their own brand of "wokewashing" as they attempt to align themselves or be in solidarity with these communities (Westervelt, 2021). A continued vigilance is needed because these "sophisticated propaganda campaigns designed to manipulate public and elite perceptions of the major oil companies are a significant barrier to meaningful climate action" (Brulle et al., 2019, p. 99). Again, these propaganda campaigns have also targeted schools for decades (Worth, 2021).

Most troubling is that climate denial impedes effective climate policy and civic action. Like other forms of denial, climate denial weakens possibilities for genuine public debate and deliberation and diminishes our capacities to identify and enact appropriate solutions. The scope of the crisis necessitates substantive debate and timely action about the ways we produce energy, grow and distribute food, and move ourselves from one place to another (Klein & Stefoff, 2021). Climate denial (whether literal, interpretive, or implicatory) breeds inaction, impugning or rejecting the need for proposals, policies, and programs to deal with the devastating and disproportionate effects of climate change. Put simply, confronting climate denial is essential to advance climate justice.

CONCLUSION

Climate-related impacts will undoubtedly worsen if denial of climate science research and denial of clear implications for what needs to be done persist. As climate leaders have been stressing (some for decades), bold, swift action is needed. The next chapter outlines a Climate Denial Inquiry Model that can help educators and students take action in confronting denial texts and contexts (and stories-we-live-by) as part of classroom practice.

CHAPTER 2

Climate Denial Inquiry Model

A two-minute video titled *Why Climate Change Is Fake News* was posted on Facebook in June 2018: https://www.facebook.com/watch/?v=854579488082979. Three years later this video had been viewed more than ten million times (or at least partially viewed; the Facebook algorithm counts unfinished views). Here is a brief excerpt from the video. The "talking point" in the excerpt refers to the 97% consensus among climate scientists about human-caused global warming.

> It's a talking point designed so you don't have to think. All scientists agree. 4 out of 5 dentists recommend this toothpaste. Why wouldn't I buy that toothpaste? Why wouldn't I believe in climate change?

The video argues that climate science is based on three falsehoods: (1) there is a 97% consensus among climate scientists about anthropogenic global warming (AGW); (2) the preceding decade is marked with record-breaking high temperatures (narrator claims these are "merely political statements"); and (3) carbon dioxide is the "control knob of the climate." A number of provocative images (e.g., Al Gore, icebergs, science labs, smokestacks, forests) with a steady, somewhat unnerving musical score accompany the narration across the 2-minute video.

Now consider a second text, a statement from the associate director of the Massachusetts Petroleum Council, taken as written testimony regarding a Massachusetts state house bill.

> Even if Massachusetts reduced its CO2 emissions to zero, it would have little impact on total U.S. emissions and no impact on global emissions, much less global temperatures. The ability of Bay Staters to heat and light their homes should not be put at risk for a policy that has zero benefits. (as cited in Lamb et al., 2020)

A third text to inspect is a McDonald's commercial that aired during the 2018 National Football League Super Bowl (just about any Big Mac commercial would work). This particular advertisement is called "Cherish" and repurposes the language of romantic commitment akin to wedding vows as a Big Mac devotional. Shuffling through several people, the words spoken are

> I renew my promise to love and cherish you. I can't help but to come back to you. I love you and only you because you have that special sauce forever. [Transition to voiceover] Rediscover your love for the Big Mac and its special sauce . . . https://www.youtube.com/watch?v=e9W1wNLbOFA

Finally, let's consider a YouTube video entitled *Nature-Based Solutions and Shell/New Energies,* which is part of Royal Dutch Shell Oil's "MakeTheFuture" campaign, and accessible at https://www.youtube.com/watch?v=p-_peqYDtoA. Less than 1 minute in length, this video flashes images that align with the following text against a backdrop of quickly paced music. We include a "/" to signal line breaks in the way the text appears in the video.

> Trees are vital in the fight against climate change / Shell is harnessing nature / Supporting reforestation efforts / Protecting forests under threat / Making it easier for customers to tackle their emissions (image of cars refueling at Shell gas station) / But this is all just one part of the solution / We aim to be a net-zero emissions energy business / by 2050 or sooner, / in step with society / So we're also investing in other areas . . . / from lower carbon biofuels and hydrogen / to electric-vehicle charging, / solar and wind power / Shell / Powering progress together / shell.com/naturebasedsolutions / #MakeTheFuture

At the end of the video the following text appears in fine print:

> It is important to note that as of May 5, 2020, Shell's operating plans and budgets do not reflect Shell's net-zero emissions ambition. Shell's aim is that in the future its operating plans and budgets will change to reflect this movement towards its new net-zero emissions ambition. However, these plans and budgets need to be in step with the movement towards a net-zero emissions economy within society and among Shell's customers.

So, how might we engage with these four texts? How can we skillfully investigate and evaluate them? We propose that a Climate Denial Inquiry Model can guide these efforts.

INQUIRY

In previous work we defined the goals and purposes of social education in terms of "inquiry-based social practices for understanding and addressing problems, especially complex, multifaceted problems" (Baildon & Damico, 2011a). We also outlined a five-step inquiry model to correspond with particular teaching and learning challenges: (1) defining purposes and launching an inquiry; (2) discerning credibility of textual sources; (3) adjudicating among conflicting claims and evidence in a range of texts, especially complex multimodal texts; (4) synthesizing findings; and (5) communicating new ideas. We carry this conception forward here. At its core, the Climate Denial Inquiry Model (CDIM) involves posing questions, locating and discerning the credibility of sources, carefully evaluating the soundness and strength of claims, including "rival hypotheses and perspectives," reasoning with evidence, and "the methodical building of evidence-based claims" (Parker, 2011, p. xiii). Overall, these processes, norms, and standards of building reliable knowledge through inquiry are fundamental to our public institutions—they are processes used by journalists, lawyers, policymakers, scientists, and academic researchers—and help inform public decision-making (Rauch, 2021). These processes, norms, and standards can also guide what happens in schools where classrooms can be communities of inquiry-based practice(s).

CLIMATE DENIAL TEXTS: SCIENCE DENIAL AND ACTION DENIAL

Given the pervasiveness of climate denial in the United States, we need a clear conception of what constitutes a climate denial text. Texts that directly deny the science and the ecological reality of our historical moment represent one type of denial text. We call these *climate science denial* texts, and the Facebook video above, with its outright dismissal of climate science, is a clear example of this denial text type.

Another type of denial text acknowledges climate science yet dismisses, obscures, or minimizes the implications of the science, thus impeding the kinds of action necessary to address the climate crisis. We call this type of text *climate action denial* because it promotes an inadequate response to the climate crisis. This term also aligns with what Cohen (2001) calls "implicatory denial": when needed action in response to facts is ignored or does not happen. Because climate action denial texts do not reject the climate science consensus, they are often more insidious and difficult to discern. The second, third, and fourth sources above are *climate action denial* texts, as the following sections will highlight. Three criteria help outline what counts as a climate action denial text. A climate action denial text

1. claims or suggests that industries or nations cannot or should not decarbonize or reduce emissions in line with IPCC goals (roughly half by 2030; net zero by 2050);
2. claims or suggests that other priorities, such as economic, political, or social goals, are more important (or separate from decarbonization efforts) or that IPCC goals must be balanced with other priorities;

Figure 2.1. Climate Denial Inquiry Model

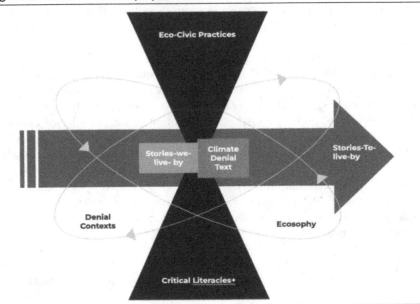

3. provides a misleading impression that a group, company, institution, or other entity is meeting IPCC goals. This is often achieved by using techniques of greenwashing, wokewashing, or paltering (equivocating or acting deceitfully). Of note, some environmental and corporate watchdog groups contend that even using the term "net zero" can be a form of denial because the term is being used to camouflage business as usual practices (Real Solutions Not "Net Zero," n.d.), especially when the goal is set decades into the future, such as pledges to be net zero by 2050.

Given the proliferation and potency of climate denial texts (science and action), we locate sources like the four texts that opened this chapter at the center of our model alongside stories-we-live-by (Figure 2.1).

STORIES-WE-LIVE-BY

Stories-we-live-by, these deep cultural assumptions about the world, can be classified as either destructive, ambivalent, or beneficial (Stibbe, 2021). Destructive stories lead to ecological harm, such as *humans are the center of existence and separate from nature*, and *the primary goal of a society is economic and technological growth without limits.* In some respects, it can be argued that these destructive stories highlight how "we all live inside" a wider story "written by coal, oil, and extractivism" (Klein & Stefoff, 2021). A related story is that freedom is solely a matter of individual freedoms (sometimes at the expense of the common good), as being free from government regulation (e.g., freedom to exploit the earth's natural resources for profit), and freedom from any other limits for that matter, often conceptualized in ways that have diminished or denied freedom to others (Monbiot, 2017; Nelson, 2021). "Freedom" is often used in marketing and propaganda campaigns, such as when the Trump administration promoted natural gas by rebranding it as "freedom gas" (Mann, 2021; O'Neill, 2019). In terms of climate denial texts, freedom can also mean the liberty to ignore climate science, to believe what you want regardless of evidence that might challenge your beliefs, and to be free of having to take any responsibility for climate change impacts. These interpretations of freedom can be called "ugly" because they are coercive and exclusionary rather than expansive (Anker, 2021). Related stories-we-live-by about technological innovation and economic growth are assumed to lead to social progress, primarily defined as material progress and expansion, increased production of goods and services, and bountiful consumerism.

DENIAL CONTEXTS

Wider contexts of denial shape the ways people engage with denial texts, how they determine their reliability, make meaning of the information they contain, develop opinions or beliefs, and make decisions about what should (or should not) be done. These contexts appear as a swirling mix in the CDIM. To understand why so many people are susceptible to climate denial texts and deny the empirical evidence and scientific consensus about whether human-induced climate change is occurring, it is important to understand these contexts and the ways they hamper our ability to collectively address climate change with required urgency.

As Chapter 3 demonstrates in greater depth, broader contexts of denial (economic, political, cultural-historical, sociopsychological, geographic, and media) are industry-driven, politically motivated, and grounded in particular historical and cultural legacies. The result is a media environment that preys on our sociopsychological and cognitive tendencies; blurs the lines between fact and opinion, real and fake news, accurate information and misinformation; and challenges our capacities to engage in essential civic practices. Understanding this web of climate denial contexts helps us recognize corporate-led efforts to distort truth and intentionally mislead and misinform the public and minimize social harm.

ECOSOPHY

Everyone is guided by sets of values and philosophical principles, including ones held about the environment. With an emphasis on ecology, Naess (1995) defines the "norms, rules, postulates, [and] value priority announcements concerning the state of affairs" that one holds to be that person's "ecosophy" (p. 8). Ecosophies evolve and can contain a civic dimension with a vision of a better society and how it might be achieved (Stibbe, 2021). An ecosophy can help confront denial by understanding the current "state of affairs" and it can point to a vision and roadmap for the future. While there is no single correct ecosophy, each can be evaluated "by whether evidence confirms or contradicts the assumptions about the state of the world that they are based on, or whether there are any internal inconsistencies" (Stibbe, 2021, p. 12).

Ayana Elizabeth Johnson and Katharine Wilkinson (2020), co-creators of The All We Can Save Project, an organization to support "transformational climate leadership, especially among Black, Indigenous, and other women of color," use aspen trees as a metaphor to articulate their own origin story as an organization, which strikes us as a working ecosophy. They write:

> Aspens are all connected, interwoven underground, supporting each other, nourishing each other, holding each other up. They are not individual trees but a single organism that must thrive as one. And that collective spirit fueled and focused the work to come. (https://www.allwecansave.earth/story)

Beach et al. (2017) outline an ecosophy that emphasizes "world citizenship, the rights and well-being of all, and the recognition of connections between the diverse members of the world family" (p. 7). Antonio Lopez (2021) outlines an ecosophy that includes advocacy for a "regenerative ecomedia system that harmonizes humans as part of the living planet for present and future generations." And Stibbe (2021) defines his ecosophy with one word, *living,* which comes with a "normative orientation towards protecting the systems that humans and other forms of life depend on for their well-being and survival" (p. 9).

These ecosophy examples resonate with us. Our ecosophy emphasizes the need to defend and protect the systems that life depends upon. This includes care and compassion for a shared environment, equitable treatment of others, and upholding human and environmental rights. Our ecosophy is also shaped by an inquiry stance and commitment to explore core questions about how we understand our relationships to each other and all species on the planet, what the goals of a society should be, and what actions need to be taken to promote justice.

To confront climate denial, we believe that it is important for educators, as well as students, to clarify their individual ecosophies and consider developing group-based ecosophies. (Working with existing classroom or school-based value, vision, or mission statements is one way to begin.) While we do not think all literacy and social studies educators would define their own ecosophies in the same way as the examples above, we anticipate there would be wide agreement about some core values, starting with the principle that all people want and deserve access to clean water to drink and clean air to breathe. This then extends to wanting people to have accessible, healthy food to eat and safe and secure, if not thriving, communities in which to live. We also wager that a great number of educators would support human and environmental rights. An ecosophy grounded firmly in these foundational values can guide efforts to confront climate denial.

CRITICAL LITERACIES+

The aims of critical literacy include an ability to critique and transform "dominant ideologies, cultures and economies, institutions, and political systems" for the purpose of understanding "how texts and discourses work, where, with what consequences, and in whose interests" (Luke, 2014, p. 22). An emphasis on the consequences of how people work with texts raises questions like, What impact in the world might responses to texts have? (Damico, Campano, & Harste, 2009).

In our work across a range of classrooms we have relied on two metaphors—excavation and elevation—to guide critical literacies (Baildon & Damico, 2011b). Excavation practices involve the careful inspection of texts (close reading, evaluating claims and evidence, etc.), while elevation practices highlight the importance of evaluating a text in terms of the larger contexts of how it was created, disseminated, and consumed and how a given text connects to or is implicated by other texts. Comber (2013) defines critical literacy with a similar excavation–elevation framing, defining critical literacy as an "evolving repertoire of practices of analysis and interrogation which move between the micro features of texts and the macro conditions of institutions, focusing upon how relations of power work through these practices" (p. 589). Anthropologist Clifford Geertz (1975) named this movement between the micro and macro a "continuous dialectical tacking between the most local of local detail and the most global of global structure in such a way as to bring both into view simultaneously" (p. 52).

Yet, we have found that the core critical literacy questions about an author's or sponsor's agenda, claims and evidence, multiple perspectives, and other considerations sometimes are not enough to fully reckon with denial texts (Damico, Baildon, & Panos, 2018; Damico & Panos, 2018). This is the reason for our designation of *critical literacies+*. We have found that working successfully with denial texts necessitates additional, focused tools. We outline three sets of these tools: FLICC, discourses of delay, and frame analysis.

FLICC

FLICC (Cook, 2020a; Cook & Jacobs, 2017; Hoofnagle, 2007; Washington & Cook, 2011) is an acronym to help readers discern the following techniques of climate denial:

- **Fake experts:** people without climate science expertise weighing in on climate change science
- **Logical fallacies:** misrepresentation, ad hominem arguments, oversimplifications, false dichotomies (e.g., the climate is always changing, so current climate change is only natural)
- **Impossible expectations:** the setting of unattainable standards as a yardstick for scientific consensus (e.g., climate models do not make predictions with absolute certainty)
- **Cherry picking:** the selection of data, quotations, or other information to support a desired view while ignoring or denying more comprehensive evidence
- **Conspiracy theories:** explanations of a situation that attribute sinister or secret motives to a group of people (e.g., climate scientists are in it for the money or to create a one-world government)

Research scientist John Cook also led the development of a mobile application called Cranky Uncle through which users can practice identifying these five science denial techniques.

Discourses of Delay and Frame Analysis

Another useful framework to help grasp how texts promulgate climate denial is to investigate "discourses of climate delay," which include:

- **redirecting responsibility**—individuals are ultimately responsible (rather than corporations or the industry) and reducing our emissions will weaken us and benefit others who don't change their ways
- **pushing nontransformative solutions**—focus on existing and new technologies, continued value of fossil fuels, and voluntary policies rather than restrictive measures
- **surrendering to the problem**—it isn't possible to mitigate climate change because changing current ways of life is impossible and we should accept our fate, which is in nature's or God's hands
- **emphasizing the downsides**—making substantial changes will be disruptive or that abandoning fossil fuels will prevent access to modern livelihoods for the global poor or will burden our own society with large costs (Lamb et al., 2020, p. 2).

Again, these discourses accept climate science but justify inaction or an inadequate response, often by directing attention to negative effects of climate policies or by escalating doubts that mitigation efforts can work (Lamb et al., 2020). In this sense, these four discourses are distinct from outright climate denial or skepticism about climate impacts, yet are "often used in combination to erode public and political support for climate policies" (Lamb et al., 2020, p. 1).

Similar to discourses of delay, Supran and Oreskes (2021) highlight the ways frame analysis can shed light on how denial discourses or frames operate in texts. These scholars demonstrate how propaganda from the fossil fuel industry—namely, ExxonMobil—has attempted to shift responsibility for global warming from fossil fuel companies to consumers. They do this by fostering dominant public narratives, such as, "we are all to blame" for global warming and "the world needs to rely on fossil fuels into the future." In particular, Supran and Oreskes describe how a "fossil fuel savior frame" functions to downplay the seriousness of climate change and justify the continued use of fossil fuels.

Particular discourses that promote this fossil fuel savior frame include:

- **technological optimism**—future technological breakthroughs will save us from climate change
- **technological shell game**—uses "strategic ambiguity" (Schneider, et al., 2016) about current or potential technologies to address climate change, such as "clean coal"
- **fossil fuel solutionism**—fossil fuels and the fossil fuel industry offer main solutions to climate change
- **individualized responsibility**—emphasizes individual rather than corporate or government responsibility
- **climate risk**—presents climate change as a potential "risk" rather than a reality
- **greenwashing**—a corporate public relations strategy to portray an ecologically responsible image
- **energy poverty**—will result if alternative energy sources are used (Supran & Oreskes, 2021).

Of note, some of these discourses are framed with a social justice argument either by issuing warnings that moving away from fossil fuels will negatively affect marginalized communities or claiming that fossil fuel companies are in solidarity with these communities. These practices can be called "wokewashing" (Westervelt, 2021).

The denial techniques of FLICC, discourses of delay, and the fossil fuel savior frame also resonate with and reinforce stories-we-live-by in compelling ways. Discourses that "push" nontransformative solutions, technology optimism, fossil fuel solutionism, and energy poverty fortify status quo ecologically destructive stories, such as, humans are the center of existence and separate from nature; consumerism is a primary pathway to happiness; and nature is a

Table 2.1. Critical literacy questions

Critical literacy questions (adapted from Baildon & Damico, 2011b)	*Critical literacy+* questions
• When, where, and why was the text produced? • What claims are being made? • What does the creator of the text want me to think, believe, or do? • What evidence is used to support claims? • What techniques does the creator use to influence me (e.g., loaded words, emotional appeals, tone, repetition)? Are they convincing? In what ways? » Look for use of loaded terms, emotive images, combinations of different modes, etc. • How might immediate and broader contexts have shaped the text's production? Consider local, national, global, historical, social, cultural, and economic forces. • What additional information is needed to understand this text?	What science denial techniques are used in this text? (FLICC) • fake experts • logical fallacies • impossible expectations • cherry picking • conspiracy theories (Cook, 2020a) What discourses of delay operate in this text? (Lamb et al., 2020) • redirecting responsibility • pushing nontransformative solutions • surrendering to the problem • emphasizing the downsides What denial frames are evident? • technological optimism • technological shell game • fossil fuel solutionism • individualized responsibility • climate risk • greenwashing • energy poverty (from Supran & Oreskes, 2021)

resource to be used or exploited. These discourses also play (or prey) on core concepts like freedom, progress, and justice to circumvent action.

Table 2.1 summarizes two sets of critical literacy questions. The column on the left is a sample of textual analysis questions. The second column points to questions and key concepts that target climate denial.

ECO-CIVIC PRACTICES: DELIBERATION, REFLEXIVITY, AND COUNTERNARRATION

Civic participation in democratic societies includes everything from voting, volunteering, and jury service to actively participating with others in public institutions (such as governing bodies, media, schools) and working with others to improve society (e.g., in political parties, civil society organizations), using methods of political organization, persuasion, protest, noncooperation, and other forms of public action. The health of a democracy depends on an informed citizenry being able to skillfully address public problems.

More broadly, civic participation means participating in communities and practices that make up everyday life and require us to act and relate to each other as ways of living together (Arendt, 1977; Dewey, 1916). For Arendt (1977) this was an ongoing activity of learning to bear with each other while striving to maintain plurality. As both a civic and learning process, it requires coming into contact with other people's thinking, understanding diverse perspectives, deliberating with others to address public problems, and subjecting prevailing views, including one's own, to critical examination.

Walter Parker (2006) offers a way to help educators understand different "purposes, problems, and possibilities" of "public discourses in schools" that has been foundational to the ways the two of us have thought about our roles as teachers in leading classroom discussions (p. 11). Parker delineates two purposes for discussion that align with different discourse structures: (1) interpretation with a seminar structure, and (2) decision-making with a deliberation structure. The goal of a seminar discussion is for students to "plumb the world deeply" as they explore challenging texts and questions where diverse interpretations are valued (Parker, 2006, p. 12). A primary goal of a deliberation is for students to evaluate alternatives related to a shared problem and then make a decision about which course of action to take, such as what to do about a school attendance policy or dress code or about a larger social problem or even an academic controversy. To promote the types of discussions needed for democratic citizenship, Parker (2006) adds three strategies—humility, caution, and reciprocity—to increase the likelihood that genuine listening can occur so "social knowledge can grow" in such a way that teachers might help "re-form" the public (Parker, 2006, p. 16).

We find Parker's distinction between a seminar discussion and a deliberation useful, and we tend to blend these discourse modes in our teaching. For example, when we emphasize the core problem or concept of *reliability,* facilitating a seminar discussion entails exploring what this concept means and what criteria might need to guide the conversation (seminar) as well as the "at the end of the day" decision-making, whether sources are reliable or not (one focus for deliberation). Because we believe it is necessary for students to put their stakes in the ground when determining source reliability, we emphasize deliberation in the CDIM.

There are also different ways to frame how people come into contact with others' views. Climate scientist Katharine Hayhoe (2021) makes a case for creating opportunities where people with varied, if not opposing, views can identify shared values or concerns. Hayhoe contends that skillful listening and question-posing can help people who might hold distinct perspectives about climate change find common ground about, for example, the value of clean water, clear air, fresh food, and the desire to ensure the health and safety of the planet for their children, grandchildren, and future generations. This perspective echoes the stance and work of longtime consumer advocate and public intellectual, Ralph Nader.

Another lens to frame dialogue across difference is agonism, or agonistic pluralism, a social and political theory that emphasizes the role that conflict plays in democratic processes as competing ideas and perspectives vie against each other. Living in a diverse and pluralistic society means that public spaces can be sites of intense conflict with little possibility of final resolution or reconciliation (Mouffe, 2018). While different or opposing beliefs and ideas may be irreconcilable at times, people with different views are considered to be adversaries whose existence is legitimate rather than enemies to be vanquished (Mouffe, 2013).

Regardless of the deliberation approach, participating in public discourses depends on being able to discern reliable, trustworthy information to make informed decisions, offer reasons to justify or defend one's thinking and actions, marshal relevant evidence to support positions, and engage in often emotionally charged argumentation with fellow citizens to determine the best courses of action.

Given that civic practices are "at heart a kind of social relation" (Burbules, 1993), an essential eco-civic practice is reflexivity, defined not as an involuntary action or performance without deliberate thought, as when a doctor taps on one's knees with a tiny rubber "reflex" hammer to assess our nerve functioning. Instead, reflexivity is a method that involves internal dialogue and grappling with factors that shape our sense-making and actions in the world, including personal, cultural, and structural factors (Archer, 2012). Reflexivity entails being willing and able to examine how our interpretations and sense-making are shaped by what we, as readers and viewers, bring to texts, including our beliefs, values, emotions, perspectives, and varied social identities or locations (race, class, gender, religion, dis/ability, sexuality, among other identity markers). How we think and act in the world is influenced but not determined by these identities or social locations, which do not operate in monolithic ways. Reflexivity involves a process of vigilant self-reflection and posing questions like: How might my beliefs, background, values or knowledge affect the way I read or interpret this source? (Baildon & Damico, 2011b).

In this sense reflexivity is both an essential critical literacy practice and a core civic practice. It calls for individuals to inspect what they bring to texts and to public issues in order to become aware of the effects their perspectives, positions, exchanges, and the language that they use might have on others, to consider how they might communicate across difference. This commitment to reflexivity helps us become more aware of the limits of our own views as we engage different perspectives and pose questions like: How might others with different backgrounds, beliefs, values, or knowledge shape the way they might engage with this source? (Baildon & Damico, 2011b). Escalating ecological, social, political, and economic challenges posed by climate change also suggests a need for "hyper-reflexivity," in which a commitment to critical literacy and the role of emotions in sense-making, or affective literacy, involves

> recognising and accepting responsibility for the harmful impacts of one's desires, investments, and emotions, and facing situations that are equivocal, frustrating, and uncomfortable without becoming overwhelmed or immobilised. Affective literacy can also be understood as the self-reflexive capacity to disarm our common defences and develop a suspicion about our own presumed innocence and the benevolence of our desires. (Stein, 2021, pp. 489–490)

As neuroscientists have been finding, emotions are core to all sense-making, inextricably tied up in thinking, rather than something separate from rationality (Barrett, 2017; Immordino-Yang & Damasio, 2007).

Melissa Gibson (2020) makes a case that deliberation approaches need to center the voices and experiences of people from the margins because a commitment to classroom deliberation alone is insufficient to help develop justice-oriented citizens. Drawing from work in critical race theory and understandings of structural, institutional, and intersectional oppression (Taylor et al., 2009), Gibson proposes *counternarration* as a way to ensure students learn to examine dominant and nondominant narratives in political and social discourse. As ensuing chapters highlight, counternarration is a critical tool to help identify destructive stories and move toward new *reliability stories-To-live-by* and *ecojustice stories-To-live-by*. To help counter forces of the climate denial machine, Indigenous narratives of place, accounts of climate-centered gender justice, as well as stories of youth-led activism can be emphasized.

Deliberation, reflexivity, and counternarration are essential tools to engage with denial texts. These approaches, working in tandem with critical literacies+, help guide the collaborative meaning-making often necessary to work through the complicated features and contexts of denial texts. Most importantly, deliberation, reflexivity, and counternarration promote dialogue across difference, as readers come to texts with varied backgrounds, experiences, beliefs, and perspectives to jointly grapple with sources, discerning their reliability, trustworthiness, and utility.

Finally, we emphasize the *eco* (or ecological) with our conception of eco-civic practices in the CDIM as a reminder to prioritize ecological issues, problems, and solutions in classroom discussions. Moreover, while the type of deliberation we explore in subsequent chapters takes place among students in a classroom or in pairs of students, there is the possibility of including some aspect of the natural world (e.g., land, a forest, animals, fish) as a silent but present dialogic participant in a deliberation (Nader, 2016a). An ecological frame, with an emphasis on relationships among organisms and their environment, is fundamentally social and relational, which is core to deliberation.

STORIES-*TO*-LIVE-BY

As described in the Introduction, this book takes aim at two kinds of stories-we-live-by. The first kind refers to dominant cultural narratives like *humans are separate from nature, the primary goal of society is economic growth without limits,* and *nature is a resource to be used and exploited.* The second kind are stories about reliability and trust, which appear to have social value but can also promote ecological harm. These *reliability stories-we-live-by* include the (1) having the "other side" represented leads to a more balanced understanding of climate change, (2) having more information or evidence is needed to support an argument, and (3) reasoning involves the evaluation of one's own identity and perspectives (Damico, Baildon, & Panos, 2018).

The core work outlined in the CDIM is to identify denial texts and discern how the techniques and tactics of denial overlap with these two kinds of ecologically destructive stories-we-live-by. With the aid of critical literacies+ and eco-civic practices, educators and students can work toward more justice-centered *reliability stories-To-live-by* and more equitable or *ecojustice stories-To-live-by.* As Chapter 6 highlights, *reliability stories-To-live-by* include: (1) climate science is a well-established knowledge base that merits public or societal trust; (2); the type and quality of evidence determines the soundness of an argument; and (3) reasoning involves the critical evaluation of varied perspectives, responsibility, complicity, and privilege. And Chapter 8 explores three *ecojustice stories-To-live-by:* (1) All life is treated with respect, care, and responsibility, especially for our most vulnerable populations and species; (2) the primary goal of society is human, ecological, and planetary well-being that comes with a recognition of limits; (3) civic engagement for the common good is necessary for more just and meaningful lives and futures.

In sum, confronting climate denial requires inquiry-based explorations of denial texts and the stories-we-live-by as they are situated in contexts of denial. Critical literacies+ and eco-civic practices are foundational to these inquiries as students learn to discern techniques, discourses, and frames of denial texts and stories-we-live-by. Now let's return to the four denial texts that opened this chapter and explore the CDIM in action with an emphasis on critical literacies+.

CDIM IN ACTION

Facebook Video

The Facebook video exemplifies *climate science denial* as it directly refutes the established scientific consensus. In terms of core literacy practices of excavation, there is a need to examine whether claims in this Facebook video are supported by evidence and to consider specific techniques used in the video, such as those used to trigger emotions in ways that will likely appeal to conservatives (and likely to anger liberals). Elevation points to larger political and economic contexts of the text's creation and possible reasons why it was viewed so many times on Facebook. Why Climate Change Is Fake News is published by Let Freedom Speak, a Facebook group that identifies itself in the "About" link, as "Conservative commentary for America's most important issues." The narrator, Mark Morano, a prominent climate change denier, is the executive director of the industry-funded Climatedepot.com, which is tied to the conservative think tank Committee for a Constructive Tomorrow (CFACT), where Morano serves as director of communications. By starting with "what the left gets wrong about climate change," the political perspective of these organizations is clear from the outset.

With sociopsychological contexts of climate denial in mind, the claims in the video that Democrats use climate science to make "political statements" and perpetuate climate change "myths," "designed so you don't have to think," can be understood through lenses of selective perception, identity-protective cognition, confirmation and disconfirmation bias (explored further in Chapter 3). While both liberals and conservatives are susceptible to these psychological tendencies, the flippant comparison of the climate science consensus to "4 out of 5 dentists recommend this toothpaste" are triggers that likely appeal to conservative viewers who already view climate change as a politically motivated issue or dismiss climate science in conspiratorial terms (e.g., as a "hoax" by scientists or liberals). Such sound bites are also likely to resonate in echo chambers in the ways that texts circulate and gain currency on social platforms like Facebook. This video likely connects with viewers who consider themselves independent-minded defenders of freedom (Let Freedom Speak), unwilling to bow to what they might perceive as challenges to their beliefs and values (as an identity threat).

This climate denial Facebook video is also financed and disseminated through a climate change "denial machine" consisting of fossil fuel industries and their well-resourced conservative foundations and organizations (Dunlap & McCright, 2011). Based on investigative reporting by DeSmog (https://www .desmog.com/committee-constructive-tomorrow/), CFACT promotes a free-market, anti-regulatory approach to environmental issues and receives the bulk of its funding from conservative foundations such as DonorsTrust ($7.8 million), described as "the dark money ATM of the conservative movement" (Kroll, 2013). As a vehicle for wealthy conservative philanthropists, DonorsTrust also contributes to climate denial groups such as the Competitive Enterprise Institute, the Heartland Institute, and the Americans for Prosperity Foundation (DeSmog, https://www.desmog.com/who-donors-trust/). The Facebook video text is situated in an economic context of organizations that are likely "to have written and disseminated texts meant to polarize the climate change issue" (Farrell, 2016, p. 92).

The climate denial machine effectively uses social media like Facebook to spread misinformation about climate science or to promote fossil fuels as part of climate change "solutions" (Kelly, 2021; Dunlap & McCright, 2011). Facebook, in particular, has spread a "staggering" amount of climate misinformation, with only a few publishers, like Townhall Media (founded by the Heritage Foundation), Media Research Center (financially supported by wealthy conservative donors and ExxonMobil), and right-wing Breitbart News, among other similar outlets responsible for 69% of the digital climate change denial content on Facebook (Buchan, 2021; Paul, 2021; Zakrzewski, 2021). Social media platforms enable the climate denial machine to connect with receptive audiences.

It is also clear that critical literacies+ and the FLICC framework in particular can help identify how climate denial works in this Facebook video. For starters, the video employs the tactic of *fake experts* as it cites Richard Tol as "one of the world's top United Nations' scientists." Tol is an economist, not a scientist and, more specifically, not a climate scientist. *Logical fallacies* also abound in this short video. There is oversimplification and misrepresentation when the narrator claims or suggests that climate scientists have been arguing that "CO2 alone controls the climate." Climate scientists are well aware that many factors affect the climate; they also know that greenhouse gas emissions linked to fossil fuel extraction and production have led to the current climate crisis. In short, climate scientists are not claiming that CO2 "alone controls the climate." The narrator also uses a "red herring" (something that distracts from the key issue at hand) by comparing a belief in climate change with a choice of toothpaste. He states: "All scientists agree. 4 out of 5 dentists recommend this toothpaste. Why wouldn't I buy that toothpaste? Why wouldn't I believe in climate change?" *Cherry picking* is yet another technique used in this video when the narrator attempts to impugn the 97% consensus among climate scientists. He states: "We find out that there are not even 97 scientists. One of the surveys was only 77 scientists, and they were anonymous scientists." The narrator cherry picks this one study while ignorning the more comprehensive and extensive findings about the consensus among climate scientists (Cook et al., 2013; Myers et al., 2021). Finally, the claims in this video that impugn the credibility of climate scientists and suggest that Democrats use climate science to make "political statements" and perpetuate climate change "myths," "designed so you don't have to think" at minimum, gesture toward *conspiracy theory* explanations that the science of climate change is a "hoax" perpetrated by groups to advance their own interests.

Statement From the Associate Director of the Massachusetts Petroleum Council

This source is straightforward and is offered here as a clear example of climate action denial in which a "discourse of delay" operates to "redirect responsibility" away from the state of Massachusetts to someone else who needs to take action before it does (Lamb et al., 2020).

McDonald's Commercial

The third text, the McDonalds' commercial, can also be framed through a climate action denial lens. Within the agricultural sector, livestock production of beef is a major source of greenhouse gas emissions (carbon dioxide and methane) and McDonalds remains one of the largest purchasers of beef in the world. While McDonalds has made strides in devising more ecologically sustainable practices, such as reducing emissions in its supply chains, the Intergovernmental Panel on Climate Change (IPCC) has warned that changes in agriculture and human diets, including scaling back meat production and consumption, is essential to curbing greenhouse gas emissions and impacts (IPCC, 2020). According to the Food and Agriculture Organization of the United Nations, the production of meat and dairy accounts for close to 15 percent of greenhouse gas emissions on Earth. People in the United States are also among the top consumers of beef in the world (Oxford Martin School, 2021).

Of note, large-scale agribusinesses and related groups and associations (e.g., Cattlemen's Association, Dairy Business) have been instrumental in promoting misinformation or distorting public opinion about the clear evidence of the environmental and climate impact of animal agriculture tied to people's continued animal-based food consumption (Stănescu, 2020)—in effect, fostering and reinforcing what Stănescu (2020) calls "meat-eating denial." To further highlight the interdependence of ecological issues, any exploration of food and agriculture must grapple with the significant global problem of soil degradation caused by pollution (agricultural, industrial, and commercial), urban expansion, overgrazing, and unsustainable agricultural practices and long-term climatic changes (Maximillian et al., 2019). Moreover, for decades the denial efforts of the industrial agricultural industry have supported the fossil fuel industry's goals to undermine climate policy and environmental regulations (Banerjee et al., 2018). While in this book we focus on the fossil fuel industry, this example highlights the intersectionality of climate-related denial (and a separate book could be dedicated to disentangling agricultural industry–led denial).

"Nature-Based Solutions and Shell" Video

Critical literacies+ are also useful to analyze the fourth example that opens this chapter, the "Nature-based Solutions and Shell" YouTube video, and its climate action denial. To begin, the frame *fossil fuel solutionism* positions the fossil fuel industry as essential to enacting solutions to climate change despite its historical and ongoing role as a leading cause of climate change. In this vein, *greenwashing* as a "discourse of delay" is also at play in this video as images of trees, forests, and nature that accompany the spoken text help advance an ecologically responsible persona for Shell and its clean energy initiatives.

This video also employs a particular misinformation technique called "paltering": when companies make claims that are true yet misleading (Cook,

2020b). Fossil fuel companies might truthfully assert that they are advancing lower carbon climate solutions, including commitments to renewable energy, yet they continue to increase fossil fuel production (Taft & Atkin, 2021). In the case of this YouTube video, while it is true that in 2020 Shell set aside $2 to $3 billion annually to promote low-carbon initiatives, this figure paled in comparison to its $17 billion commitment to its fossil fuels operations (Carrington, 2021a). Paltering can be difficult to detect without an examination or understanding of the full context of a company's or the industry's activities (Cook, 2020b). In addition, in the final moments of the video it is easy to overlook the fine print that states Shell's current budget plans do not align with a net-zero commitment that was stated earlier in the video.

One can anticipate that corporate greenwashing tactics and related techniques will continue to evolve in response to growing resistance from citizens and varied environmental groups. For example, during the run-up to the October 2021 congressional hearing with CEOs of ExxonMobil, BP America, Chevron, and Shell, along with the heads of the U.S. Chamber of Congress and the American Petroleum Institute, there was a barrage of industry propaganda targeting lawmakers, lobbyists, policy professionals, and industry leaders (Taft & Atkin, 2021). This targeted disinformation campaign was designed to look like original reporting in popular newsletters covering the hearings—Punchbowl, Axios Generate, and POLITICO Morning Energy (Taft & Atkin, 2021). In the month leading up to the hearing, 63% of Punchbowl newsletters were sponsored by fossil fuel interests while 100% of POLITICO's Morning Energy newsletters and 62% of Axios Generate newsletters were sponsored by fossil fuel interests, all significant increases from previous months (Taft & Atkin, 2021).

With all four of these information sources, the eco-civic practices of deliberation, reflexivity, and counternarration are instrumental in guiding inquiry-based investigations. Without an opportunity to engage with others in these practices, students with potentially varied perspectives might not modify or deepen their understanding of these climate denial sources.

CONCLUSION

Climate denial in the United States is elaborate, pervasive, and pernicious. The Climate Denial Inquiry Model can help educators grapple with the complex components of this denial: Critical literacies+ and eco-civic practices (deliberation, reflexivity, counternarration) help students identify denial texts and discern stories-we-live-by that are situated in and shaped by intersecting contexts of denial. We examine these "web of denial" contexts in the next chapter. The results of all of these efforts can lead to enacting and embracing two kinds of justice-centered *stories-To-live-by* that are tied to how we collectively want to understand reliability and trust (*reliability stories-To-live-by*) and how we define a different set of stories we might rather cultivate (*ecojustice stories-To-live-by*).

CHAPTER 3

Web of Climate Denial

Equipped now with the Climate Denial Inquiry Model (CDIM), how might educators begin confronting climate denial in their own teaching and learning spaces? We suggest starting with what we see as a foundational inquiry question:

What best explains the range of climate change beliefs in the United States despite the overwhelming scientific evidence and consensus about the existence and causes of climate change?

Finding the right content angle or subject matter focus to draw students into a topic and enable them to explore its productive tensions or dimensions is a key starting point for designing inquiry lessons (Swan et al., 2018). For us, this inquiry question meets several important criteria: It is *authentic* (has real-world interest and application), can *engage students* and *invite multiple perspectives*, can be *investigated empirically*, and *requires an evidence-based explanation*.

Elsewhere we have outlined specific ways teachers can design a classroom activity based on this inquiry question: how to select sources to investigate the question, design tools to guide careful reading and analysis of sources, and create a culminating task where students develop evidence-based explanations to answer this question (Damico & Baildon, 2015). The activity is aligned with the College, Career, and Civic Life (C3) Framework for Social Studies State Standards (NCSS, 2013) and supports inquiry as a process for students to investigate significant questions about their world, use key disciplinary perspectives, analyze a range of sources, and develop their own explanations to address key problems in the world.

In this chapter we take a more elevated view to highlight how this and other inquiry questions about climate change are entangled in a "web of denial" with six intersecting contexts: political, economic, cultural-historical, sociopsychological, media, and geographic. While these contexts connect to content areas or subject matter foci in social studies (civics, economics, history, psychology, sociology, geography), we think it is more useful to identify these contexts as overlapping, mutually reinforcing, and fundamentally inter- or multidisciplinary. More fully understanding how these contexts function is key for teachers and students to confront climate denial texts with the CDIM. In this sense, these contexts represent core social studies content that, in turn, can support inquiry-based investigations with students.

CLIMATE CHANGE DENIAL MACHINE

Dunlap and McCright (2011) outline key components of a "climate change denial machine" led by the fossil fuel industry (e.g., ExxonMobil, Peabody Coal, American Petroleum Institute, Edison Electric Institute); organizations representing the interests of corporate America (e.g., U.S. Chamber of Commerce, National Association of Manufacturers, National Mining Association); and well-financed conservative foundations (e.g., Koch, Scaife, and Mercer controlled family foundations). These industry leaders have financed conservative think tanks that have promoted denial, such as the American Enterprise Institute, the Competitive Enterprise Institute, the Heartland Institute, and the Heritage Foundation.

This network of corporate and political leadership then shields their climate change denial activities through front groups that promote industry agendas, challenge climate science, and lobby against climate policymaking. Some front groups noted by Dunlap and McCright include the Global Climate Coalition, the Information Council for the Environment, Center for Energy and Economic Development, the Greening Earth Society, and the Cooler Heads Coalition. Despite their seemingly innocuous or proenvironment names, these groups have aggressively worked to oppose climate science and policy by hiring contrarian scientists to manufacture uncertainty about climate science and by spreading denialist messaging through conservative media outlets, political activism, and lobbying. This denialist messaging is often accompanied by powerful cultural-historical narratives of continued economic growth, jobs, and prosperity that are very appealing to people in the United States and elsewhere with promises of economic development based on fossil fuels.

Dunlap and McCright (2011) also highlight the role of astroturf groups and campaigns that appear as spontaneous, popular grassroots efforts yet are generated and financed by industry, think tanks, or front groups. Groups such as Americans for Progress, FreedomWorks, Americans for Balanced Energy Choices, and Energy Citizens hold rallies, distribute denial literature, and lobby to promote climate science denial and oppose climate policy. These highly organized, well-financed, and institutionalized efforts use conservative media and social media to create an echo chamber of climate change denial that has "had a profound effect on the way in which climate change is perceived, discussed, and increasingly debated" (Dunlap & McCright, 2011, p. 156).

Public relations (PR) firms have also played a significant yet often overlooked role in advancing climate denial. Brulle and Werthman (2021) demonstrate how leading PR firms have shaped public discourse about climate change by coining and strategically promoting phrases like "coal country," "clean coal," "carbon footprint," and "renewable natural gas." The sizable impact of PR "information and influence campaigns" highlights how PR companies are not "generic organizations" but "active agents" in the "conceptualization, design, and execution of communications and political campaigns" through core strategies of corporate image promotion (such as greenwashing), third-party mobilization (to carry out campaigns), and delegitimization of opposition and the tactics of paid media campaigns, earned media placements, grassroots rallies/events, and social media campaigns (Brulle & Werthman, 2021, p. 7). Public relations efforts can also intersect with corporate lobbying tactics about climate change that result in a type of smokescreen over environmentally destructive practices (Michaels & Ainger, 2020). Of note, there are also initiatives to direct public relations and advertising professionals away from the fossil fuel industry. The Clean Creatives Pledge (www.cleancreatives.org) is one effort in which agencies and creatives refuse to take on any contracts with fossil fuel companies, front groups, or trade associations.

An understanding of the climate change denial machine points to how economic, political, cultural-historical, sociopsychological, media, and geographic dimensions of climate denial are interrelated and work together to undermine climate science and stymie climate action.

ECONOMIC CONTEXTS

Capitalism, based on private ownership of property and the means of production, competitive markets for goods and services, and the profit motive promotes economic development and expansion (through new markets, enhanced production capacities, discovery of new resources, new forms of labor, etc.) and ensures that physical infrastructure for production and consumption (e.g., transport and communications systems, education, research) supports ongoing development, expansion, and profits (Harvey, 2004). Capitalism is premised on a story that perpetual economic growth and the relentless accumulation of wealth is the goal of individuals and society (Jackson, 2021; Stibbe, 2021). It is a story that also views nature as a resource to satisfy human desires.

The standard measure of capitalist economic performance is Gross Domestic Product (GDP), the sum value of goods and services produced in a nation. However, GDP fails to account for other significant economic factors, such as levels of employment, economic security, inequality, well-being, or social progress (Stiglitz et al., 2019). The continual drive for expansion and profit has also created externalities that aren't measured, such as the production of goods people don't need, waste, and environmental pollution (Fairbrother, 2016). The instrumental logic of limitless growth, profit maximization, GDP, and consumerism regardless of limits has degraded nature, exhausted labor, and relied on markets that are "unable to take qualitative, ethical, social, human, or natural values into account" (Löwy, 2015, p. 51). Denial is central to the economic model of capitalism—a denial of how limited and finite the world is, of social justice, and of the interconnected nature of our lives and our relationship to nature (Pope Francis, 2015).

Climate change denial is part of broader economic patterns of profit-motivated denial that have existed in American society for centuries (Freese, 2020). There have been industry-led campaigns to deny well-established evidence about the social harms of the slave trade; exposure to radium; the impact of leaded gasoline and tobacco use; and the environmental impacts of acid rain, chlorofluorocarbons (CFCs) and ozone layer depletion, as well as dichloro di-phenyl trichloroethane (DDT) (Freese, 2020; Oreskes & Conway, 2010).

Fossil fuel corporations have generally received strong bipartisan support in U.S. economic and energy policy. For example, President Obama's administration opened up millions of acres for gas and oil exploration, increased offshore drilling, and added new gas and oil pipelines across the United States (Sirota, 2021). Democrats and Republicans alike have long supported fossil fuel subsidies intended to keep energy prices low for consumers, which has made alternative energy sources less affordable in comparison. Recent energy policies have resulted in U.S. companies overtaking Saudi Arabia in oil production and exports since 2018, while energy insecurity (the inability of consumers to pay energy bills) remains a persistent issue (Chuvakhina et al., 2021; Graff & Carley, 2020).

Fossil fuel companies have also received strong, steady support from the world's largest banks. While efforts to get organizations, institutions, and other entities to divest from fossil fuels continue to be successful (e.g., Rockefeller Brothers Fund, Stanford University, Harvard University, World Council of Churches, and Ireland all have committed to divesting), big Wall Street banks remain committed to oil and gas investments, with JP Morgan Chase leading the way (McKibben, 2021). The $7.4 trillion private equity industry has also contributed hundreds of billions of dollars to fossil fuel companies, which has significant and long-lasting impacts with communities of color disproportionately shouldering the harm (Giachino & Mehta-Neugebauer, 2021). Not surprisingly perhaps, investigative reports have also demonstrated links between the world's top insurance companies and the fossil fuel industry (Sherrington, 2021).

Björnberg and colleagues (2017) highlight empirical studies from several policy fields that demonstrate the ways science denial is promoted by actors with considerable economic and political capital. The tobacco industry's persistent denial of the harmful effects of its products offers a textbook case of a well-executed game plan to misinform the public that was well documented within the industry itself (Oreskes & Conway, 2010). Inside Climate News, a Pulitzer Prize-winning, nonprofit and nonpartisan news organization, has similarly found that ExxonMobil has worked at the forefront of climate denial for decades. Based on internal company documents, interviews with former company employees, and other evidence, there is a well-documented history of ExxonMobil intentionally sowing doubt and spreading misinformation about climate change. Climate science denial has been "by far the most coordinated and well-moneyed form of science denial, constituting the backbone of the opposition to environmentalism and environmental science in general" (Björnberg et al., 2017, p. 235).

Transnational petro-capitalism, driven by the leading oil, gas, and petrochemical conglomerates around the world (e.g., ExxonMobil, Chevron, PetroChina, Saudi Aramco, Royal Dutch Shell, and BP), capitalizes on the production, distribution, and consumption of fossil fuels and petroleum-based products, significantly shaping the globalized economy, national political systems, consumer culture, and geopolitical affairs (Rogers et al., 2013). These corporations have played an outsized role in the web of climate denial and climate inaction. They have hired scientists to challenge the scientific evidence of climate change as well as economic consultants to help negate, dilute, or delay national and international climate policy (Franta, 2021).

In sum, major fossil fuel companies have used their vast financial resources to control the narrative that the goal of society is perpetual economic growth dependent on fossil fuels. The scientific evidence for climate change was irrefutable by the 1980s and initially supported by industry-sponsored scientific research at the time (Freese, 2020). However, the industry narrative quickly changed in response to climate action initiated by the Montreal Protocol in 1987, the establishment of the IPCC under United Nations auspices in 1988, the first IPCC report on the climate in 1990, and the Rio Earth Summit among world leaders in Rio in 1992. Fossil fuel industry opposition and denial campaigns were mobilized as the scientific evidence mounted (Freese, 2020). In response, the fossil fuel industry aimed to discredit climate science while promoting their own pro-industry narratives. Major oil and gas companies have not only funded climate denial. In terms of their own advertising campaigns, these companies have relentlessly pushed pro-fossil fuel propaganda to further remind and convince people about how essential fossil fuels are to society (Brulle et al., 2019). From the industry's perspective, there is "no short-term economic incentive for stuffing the fossil fuel genie back into its bottle . . . fiscal cycles turn much faster than biogeochemical cycles, and there is no industrial profit to be made in pushing a Story of Less" (Jahren, 2020, p. 138).

POLITICAL CONTEXTS

Political power is fundamentally about the ability of one group of actors to shape behaviors in society, determine the distribution of limited resources, and set political agendas that often benefit some while disadvantaging others

(Hancock & Vivoda, 2014). Since the mid-19th century, politics in the United States coevolved with the fossil fuel–based economic contexts of industrial capitalism and the rise of oil barons like John D. Rockefeller, Andrew Mellon, and J. Paul Getty, who were able to influence the government to create monopolies and amass tremendous wealth (Yergin, 1990). In recent decades this evolution is marked by a sharp rise in corporate power and a decline in regulatory controls, resulting in the triumph of neoliberalism as national and global political projects (Harvey, 2006). The result is that powerful corporations wield considerable political influence to promote their own agendas.

More-recent corporate influence in politics has been carried out by a network of extremely wealthy coal, oil, and gas magnates who have formed the nucleus of the Koch donor network (Mayer, 2016). In *Dark Money: The Hidden History of the Billionaires Behind the Rise of the Radical Right*, Jane Mayer (2016) outlines how the grandchildren of oil barons, pioneers of fracking, CEOs of energy companies, heirs to oil-drilling fortunes, and owners of pipeline, drilling equipment, and oil service companies have provided funding to fight climate reform. In particular, Charles and David Koch, the Bradley Foundation, and the Scaife and DeVos families were among some of the largest donors financing what Brulle (2014) has called the climate change countermovement. Using donor-advised funds like Donors Trust and Donors Capital Fund, wealthy donors spent over half a billion dollars on lobbying efforts from 2003 to 2010 to influence policymaking and mislead the American public about the threat of climate change (Brulle, 2014; Mayer, 2016).

Corporate influence has led to the capture of public institutions, such as our government agencies, media, judicial system, and schools, in ways that affect public policy, influence government regulation or intervention, and limit the role of citizens in democratic decision-making (Bó, 2006; Nyberg, 2021). Corporate political influence is carried out through political campaign contributions and lobbying, funding political action committees (PACs) and super PACs, commissioning policy research, spending on public relations and advocacy advertising, and creating industry associations and political alliances to help corporations realize private gains and influence policy decisions favorable to firms (Dunlap & McCright, 2011; Nyberg, 2021). Such efforts corrupt democratic processes by excluding citizen interests, voices, and representation from public deliberation (Nyberg, 2021).

An example of fossil fuel industry capture of the executive branch during the Trump administration included the appointments of people with ties to the fossil fuel industry to top positions in the Department of the Interior (Montana Congressman Ryan Zinke), the Environmental Protection Agency (Attorney General from Oklahoma Scott Pruitt), the Department of Energy (former Texas Governor Rick Perry), and Secretary of State (former CEO of ExxonMobil Rex Tillerson) (Roberts, 2017). This enabled key government agencies to roll back regulations, cut agency budgets, staff these agencies with people committed to fossil fuel company agendas, and promote industry-friendly policy and climate inaction. However, while policies and practices of the Obama administration did not match the scale and impact of the Trump administration's climate action denial, the Obama administration also supported fossil fuel industry interests—for example, signing a bipartisan bill ending the ban on crude oil exports instituted during the 1970s oil crisis, which was a result of intense lobbying efforts by the American Petroleum Institute and other industry groups (Aronoff, 2020).

The 2010 Supreme Court decision *Citizens United v. Federal Election Commission*, 558 U.S. 310, gave corporations and wealthy elites the power to spend unlimited amounts of money through campaign financing, special interest groups, lobbying efforts, and media campaigns to shape the ideological agendas of political parties and influence public policy. Empirical studies have found that the Citizens United "money as speech" decision lifting spending restrictions on electioneering communications has led to more ideologically conservative legislatures and that the more contributions members of Congress received from oil and gas companies supporting their reelection, the more likely they were to vote against environmental policies (Abdul-Razzak et al., 2019; Goldberg et al., 2020).

According to the Center for American Progress, the 117th Congress (elected during the 2020 election) includes 109 representatives in the House and 30 senators who refuse to acknowledge the scientific evidence of human-caused climate change. These 139 elected officials have made statements casting doubt on the established scientific consensus about climate change and have received more than $61 million in lifetime contributions from the coal, oil, and gas industries (Drennen & Hardin, 2021). Top recipients of fossil fuel industry money include prominent members of Congress, such as Republicans Mitch McConnell (over $3.5 million), Jim Inhofe (over $2.5 million), and Ted Cruz (almost $4 million; Drennen & Hardin, 2021), while Democrat Joe Manchin received $400,000 from fossil fuel industry donations and has made over $5.2 million over the past 10 years from a coal and energy resource company he founded (Hirschfeld, 2021). The money flow between fossil fuel companies and elected representatives, of course, is a two-way street. According to a report from the International Monetary Fund, in 2020, fossil fuel subsidies in the United States amounted to $660 billion (yes, billion with a "b") while globally, the figure was $5.9 trillion (Parry et al., 2021), which amounts to $11 million a minute (Carrington, 2021b). (Figures include explicit price subsidies along with implicit environmental, tax, and health subsidies.)

Some time ago, political theorist C. Wright Mills (1956) warned about the alliance of corporate and political elites to diminish politics as genuine public debate and democratic decision-making. The cozy relationships between corporate officials and political leaders has been well-documented more recently—for example, in episodes of the Ralph Nader Radio Hour (https://ralphnaderradiohour.com) and Alex Gibney's Netflix series, *Dirty Money.* In *The Chickenshit Club: Why the Justice Department Fails to Prosecute Executives,* the journalist Jesse Eisinger (2017) notes the "revolving door" between government officials (e.g., from the Justice Department) and corporate executives that results in corporate malfeasance often going unpunished. Whether it is ExxonMobil's intentional public misinformation campaigns about climate change, the role of utilities companies spending lavishly to thwart renewable energy policy, or all-too-familiar industrial-environmental disasters, the corporate-political establishment tends to enable continued assaults on our environment and thwart effective climate policy.

Political power shapes what gets said and heard in the public sphere—climate denial can be bought, sold, and amplified to crowd out climate science or at least sow enough doubt so that less than a quarter of Americans say they understand the scientific consensus. Wealthy elites have been able to "buy an infrastructure of persuasion not available to others" (Monbiot, 2017, p. 147). Politically motivated communication drowns out verifiable information as well as a range of perspectives, leading to a decline in democratic discourse, making it more difficult to effectively address significant public problems, such as climate change (Kavanagh & Rich, 2018).

CULTURAL-HISTORICAL CONTEXTS

Stories structure our language and shape the ways we think about ourselves, our relationships to others, and our environment in ways that tend to be far more important than facts or evidence in helping us make sense of and navigate the world (Monbiot, 2017). Powerful cultural narratives include notions of the earth as existing for the extraction of natural resources to satisfy human needs and wants (Klein, 2014); a story of human empire and domination that brings endless prosperity and security (Korten, 2007); a story of "business as usual" encouraging us to continue producing and consuming more than we already do (Macy & Johnstone, 2012); a story of fierce competition with winners and losers that often blames the "losers" for social problems (Monbiot, 2017); and a zero-sum story that undermines the potential for solidarity across groups (McGhee, 2021).

Consider two interrelated stories-we-live-by that permeate our culture: (1) humans are separate from nature, and (2) consumerism, or the acquisition of goods, services, and experiences, is a primary pathway to happiness and fulfillment (Damico et al., 2020). Together these stories are destructive because they perpetuate ecological injustice and "lead to both alienation from life and environmental destruction" (Stibbe, 2021, p. 3). These propositions are untenable—earth's finite resources cannot support an indefinite expansion of industrial civilization (Lasch, 1991). These stories promote cultural propositions that are a denial of the limits of nature and the human condition (Baildon & Damico, 2019; Lasch, 1979).

Elements of these stories, in neoliberal form, include notions of competition, free markets, anti-regulation and limited government, privatization and unfettered individualism, and freedom to expand markets, plunder natural resources, and exploit labor. Neoliberal notions of freedom and progress as endless entrepreneurial advancement and personal gain is an extension of hypercapitalism leading to environmental, social, and personal exploitation and exhaustion (Han, 2015). These legacies go back to early capitalism and colonialism and provide a "logic of domination" (Warren, 1990, p. 125) with hierarchized binaries that inform how we conceptualize social relationships and our relationship with nature (Wolfmeyer et al., 2017). These deeply rooted cultural narratives deny forms of knowledge, relationships, values, and social practices premised on other cultural, epistemic, and ecological paradigms. Traditional, Indigenous, and local knowledge systems, for example, have been denigrated, devalued, and destroyed.

These stories also reflect corporate public relations and media efforts to frame corporate agendas as in the public interest through a model of progress based on endless economic growth, technological advancement, free markets, consumerism, and the control of nature. This messaging is promoted in political agendas and policymaking, framing how public issues are communicated and debated. For Bowers (1997) this framework is part of a broader culture of denial characterized by an anthropocentric view of the world and of the individual as the basic social unit, assumptions about development and progress as mainly economic and technological, and capitalist views of the world as the most legitimate explanations of the world and our role in it.

Instead of simply denying basic facts and implications, alternative stories and "facts" are presented as well. Denial is suffused with conspiratorial narratives (e.g., Lewandowsky et al., 2015; McKewon, 2012). For example, accusations of conspiracies of a corrupt peer review process within the IPCC were reported in the Wall Street Journal as early as 1996 to discredit climate science (Lewandowsky et al., 2015; Oreskes & Conway, 2010). Lewandosky and colleagues (2015) present a whole list of climate denial books that contain conspiracist themes in their titles, such as *The Greatest Hoax, Climate of Corruption, Climategate,* and *The Deniers—The World-Renowned Scientists Who Stood Up Against Climate Warming, Political Persecution and Fraud.* Conspiracy theory, denial, and a paranoid style in politics have been recurrent forces in American public life, often conveyed by narratives of perceived (or politically manufactured) threat, sinister intent, suspicion, blame, and victimization (Cook, 2020b; Hofstadter, 1964).

The dominant cultural narratives or stories-we-live-by that promote denial also tend to be nonpartisan, operating powerfully across political lines. Neoliberalism, for example, has been the dominant economic and political storyline of both conservatives and liberals since the 1970s. Both liberals and conservatives have essentially supported corporate global capitalism and energy policies that have resulted in increases in carbon emissions and various forms of climate denial (Fraser, 2017). While some Democrats have led efforts to combat climate change (former Vice President Al Gore is one prominent example), Democrats have been complicit in political discourses that have effectively delayed climate action, such as those that advocate mainly symbolic ("all talk, little action"), voluntary, or nontransformative solutions; promote stories of technological optimism that future innovation will address climate change; and push narratives of individual rather than corporate responsibility (Lamb et al., 2020). Liberals and conservatives have both participated in forms of implicatory denial (Cohen, 2001) by refusing to fully acknowledge or act upon the personal, moral, and political implications of climate science. Aronoff (2021) points to particular U.S. aspects of climate denial, which include "one part paranoid anticommunism, one part corporate capture of politics, and a healthy dose of manifest destiny, funded by the executives with the most to lose in the transition to a low-carbon world." In the United States there is now greater acceptance of climate realities and push for needed action in the current Democratic party, while the Republican party remains more rigidly opposed ideologically to climate action.

SOCIOPSYCHOLOGICAL CONTEXTS

Denial is a psychological defense mechanism used to protect the ego against painful or uncomfortable social experiences and the ideas and emotions they may evoke (Freud, 1936). It serves as a coping mechanism that allows "ordinary life, in some form, to continue" in the face of challenging realities (Randall, 2005, p. 167). Denial helps us manage "difficult knowledge" which might disrupt our sense of self and what we value, "know," or believe to be true (Britzman, 1998; Garrett, 2017). Denial encourages us to hold onto stories-we-live-by even if they have become destructive.

Denial is supported by three interrelated social psychology processes in which information that doesn't align with existing emotions and beliefs is rejected: (1) *politically motivated reasoning*, in which one's political views or group membership affect cognitive processes; (2) *confirmation bias*, the tendency to find and interpret information in ways that confirm existing beliefs; and (3) *identity-protective cognition*, when people trust and process information based on their identity and group membership. These processes, which are "fundamental to our basic cognitive architecture" (Kraft et al., 2014, p. 130–131), can lead individuals to deny evidence or information that goes against desired beliefs. Motivated reasoning helps us protect our sense of identity and our identity group's status and beliefs (identity-protective cognition)—it is suffused with emotion and often triggered automatically and unconsciously by stimuli (Kahan, 2017; Kahneman, 2011; Kraft et al., 2014). Motivated reasoning can lead to confirmation bias—the desire to find more satisfying confirmatory evidence to support existing beliefs (Shermer, 2011). Identity-protective cognition can lead to a "backfire effect" in which initial beliefs are reinforced when challenged (Nyhan & Reifler, 2010).

The psychological and emotional impact of climate-related concerns is also evident. An international study of climate anxiety among 16–25-year-olds highlighted significant levels of climate-related distress globally, with many young people experiencing a range of emotions such as worry, fear, anger, grief, despair, guilt, and betrayal, as well as hope (Jensen, 2019; Marks et al., 2021; Ojala, 2017). The refusal to acknowledge and act upon the personal, moral, and political implications of climate change may account for why 40% of the American population feel helpless about climate change (Leiserowitz et al., 2020). Denial as a defense mechanism is one way to manage the stark realities of climate change and the emotions they evoke.

Social norms and practices also contribute to climate change denial. Trust is a fundamental social, emotional, and rational component of what we know and believe (Druckman & McGrath, 2019; Shapin, 1994). Who we trust as sources of information affects how we understand climate change, view our personal responsibility toward it, and decide which policies we support (Bodor et al, 2020). Since we generally lack the specialized knowledge necessary to verify or fact-check complex information, we rely on others for much of this knowledge. We sometimes trust information even when we have reason to doubt it (Sunstein, 2021). Even lies, alternative facts, and misinformation can be accepted if somewhat plausible (i.e., "it might be true, maybe it is true") or if believing them allows us "to remain part of a tribe that accepts us and makes us feel safe" (Hayhoe, 2021, p. 7).

Finally, beliefs are more easily instilled when a person's cognitive resources are depleted; when people are confronted with conflicting information and repeated demands on their attention, their cognitive capacity to reject doubtful propositions (i.e., misinformation) is greatly reduced. Gilbert (1991) argues that "the fatigued or distracted individual" can be "especially susceptible to persuasion" (p. 111). A sociopsychological environment of 24/7 news and social media, endless distractions, conflicting accounts and evidence, denial texts, and claims of fake news contributes to this reality.

MEDIA CONTEXTS

The industry-led, political, cultural-historical, and sociopsychological dimensions of denial outlined above are amplified by our media ecosystem. For example, during the 2020 election and largely in response to presidential candidate Joe Biden's climate plan, 25 oil and gas companies and advocacy groups spent $9.6 million to share Facebook ads marketing fossil fuels or promoting fossil fuels as part of climate change "solutions" (Kelly, 2021). With ExxonMobil the top spender (over $5 million), followed by fossil fuel industry trade association the American Petroleum Institute ($2.97 million), these ads were viewed over 431 million times by Facebook users (InfluenceMap, 2021).

In their review of literature on the spread of climate misinformation (misleading information regardless of intent to deceive), Treen et al. (2020) argue that social media has made the diffusion of misinformation easier and faster. They suggest a contagion model of spread in which the "infection" of climate denial or doubt spreads to people who are "susceptible" because of their existing beliefs, thus creating an epidemic of misinformation and confusion about climate science. There are interconnected characteristics of online social networks and cultural, epistemic, and psychological factors that make some social media users particularly susceptible to the contagious consumption, acceptance, and spread of climate misinformation (Treen, et al., 2020). Online networks make it possible for public relations departments or social influencers to use techniques, discourses, and frames that "connect" to audiences in different ways (e.g., through appeals to logic, values, emotions, ideology). In particular, disinformation campaigns appeal to people's sense of identity, their group loyalties, and perceived identity or group threats. This human and technological brew contributes to public misunderstanding of the scientific consensus about global warming.

While there is a history of right-wing media activists who have understood how to spread ideas through a network of ideological media sources (including conservative publishing houses, radio programs, magazines, and political campaign funds) starting in the 1940s and 1950s, the rise of Fox News and media stars like Rush Limbaugh and Sean Hannity offered broader platforms for ideologically driven information (Hemmer, 2016; Lepore, 2021). Studies have found that conservative media are now further to the right and that conservative readers and viewers are more likely to visit fake news sites, be exposed to conspiracy theories, and to believe them than liberals (Guess et al., 2020; Rauch, 2021; van der Linden et al., 2021).

Conservative news outlets, however, are not the only purveyors of climate denial. The Cable News Network (CNN), often viewed as the political left's counterpart to Fox News, has played a role promoting denial. For example, in the 1-week period after record-setting high temperatures of 2015 and surprisingly high temperatures in February 2016, CNN aired almost five times as much advertising from the oil industry compared to its climate change–related coverage (Kalhoefer, 2016). This gross disparity did not account for the dozens of non-energy-related ads from the billionaire Koch brothers that were placed on CNN during this period to boost the image of Koch Industries.

A more comprehensive analysis across a longer trajectory also challenges common perceptions about conservative news sources as the primary provider of misinformation coming from major climate denial groups. Based on a 22-year time frame (1996–2017), over half the coverage from these groups was carried out in the *Washington Post* (32%) and *New York Times* (19%), while just 15% of the coverage was carried out in two favored conservative outlets, Fox News (14%) and the *Wall Street Journal* (1%) (Boykoff & Farrell, 2020). Another source of online climate misinformation is generated by bots, with about a quarter of all tweets about climate produced by bots on an average day, which distorts online discourse by including far more climate change denial than might otherwise be found online (Mann, 2021; Milman, 2020).

The emergence of platform capitalism (Srnicek, 2017) and the ubiquity of corporatized digital infrastructures that enable groups of people to interact (such as customers and advertisers, producers and suppliers, educators and students, and "friends") also shapes the public sphere and everyday life. Digitized public sphere platforms, based on technical and business models of social media, have normalized the use of algorithms and datafication as a new paradigm in society (van Dijck, 2014). Deep learning algorithms crunch data and perform computations to make "inferences" about people's shopping, political, and informational preferences to then feed people the same type of information they sought in past online searches and interactions (Tufekci, 2019). Profit-driven digital platforms essentially serve as confirmation bias platforms that reinforce cognitive biases and existing beliefs and validate dominant cultural assumptions and narratives. In sum, Treen, Williams, and O'Neill (2020, p. 1) argue that

> Climate change misinformation is closely linked to climate change skepticism, denial, and contrarianism. A network of actors are involved in financing, producing, and amplifying misinformation. Once in the public domain, characteristics of online social networks, such as homophily, polarization, and echo chambers—characteristics also found in climate change debate—provide fertile ground for misinformation to spread. Underlying belief systems and social norms, as well as psychological heuristics such as confirmation bias, are further factors which contribute to the spread of misinformation.

The algorithmic amplification and dissemination of misinformation can intensify sectarian views and create echo chambers of misinformation to not only splinter any hope for a shared sense of reality (Zuboff, 2021); it has also led to many embracing an outright denial of reality. Lies, produced through deliberate disinformation campaigns and emotional appeals, seeded on social media platforms and amplified through algorithms, become the basis of a different reality. Disinformation and denial is now magnified by our communications revolution—velocity of their spread; their virality (which challenges both careful, critical thinking and democratic deliberation); their anonymity (which enables unaccountable antidemocratic speech); the homophily of like-minded individuals that offers safe havens for the propagation of conspiracy theories; and the monopolies and lack of regulation that social media platforms have in the information ecosystem—all of these contribute to the complex social problem of denial (Persily, 2019).

As an example of the amplification and viral spread of climate disinformation, consider what happened on social media (Facebook, Twitter, YouTube) during the February 2021 power outages in Texas. An analysis of Facebook posts about the outages showed that 99% of climate disinformation related to this event was not fact-checked (Johnson, 2021). The false narrative that the blackouts were caused by failed wind turbines showed how disinformation, although debunked by Politifact, *USA Today*, Reuters, and other media outlets, spread on social media so that falsehoods became talking points for prominent media personalities (e.g., Tucker Carlson and Sean Hannity) and politicians (e.g., Texas Governor Greg Abbott, Texas Senator John Cornyn, Texas Representative Dan Crenshaw, Ohio Representative Jim Jordan) within 4 days (Friends of the Earth, 2021; Shepherd, 2021). The report reveals how quickly "right-wing extremists and fossil fuel interests weaponized social media to deride climate solutions" by expanding into mainstream media and the political arena "false talking points for politicians to blame renewable power and climate solutions at large for the failures of fossil fuels in Texas" (p. 2). Attempts to rein in climate denial on the Internet can also be challenging. In late 2021, Google, for example, announced it would no longer place ads on any sites that deny climate change science, including ads that refer to global warming as a hoax, and that state climate change is human-caused. Yet there was evidence that denial-laced ads continued (Tabuchi, 2021).

The Center for Humane Technology (2021) offers a ledger of harms based on scientific research that includes the viral spread of misinformation, conspiracy theories, and fake news—highlighting how social media can undermine our ability to focus on, think about, understand, and act on complex social issues (like climate change). The center cites research, including work in neuroscience, that suggests social media drains our attention away from important matters, increases the stress we feel, and promotes more extreme views. Research also highlights that sustained disinformation campaigns, such as those promoted by industry and particular media outlets, can dilute, distract from, and deny the realities of different forms of injustice. Digital platforms not only distance people from others and from nature, but their effects can contribute to people feeling disengaged, doubtful, and dismissive of climate science.

GEOGRAPHIC CONTEXTS

Spatial narratives shape how we think about the world (Schlemper et al., 2018), and the Global North and South divide is one way to frame climate change impacts and shape how global warming might be addressed (Muller, 2020). The global divide narrative helps to name practices of colonialism and imperialism, in which powerful nations use their economic, political, and military power to extract resources, open markets for their goods and services, or exert their dominance over weaker nations for their own strategic advantage. Historically, the headquarters of most transnational petrochemical corporations have been in Global North nations (mainly in North America, Europe, the Middle East, and East Asia) where more than half of global industrial greenhouse gas emissions were traced to just 25 corporate and state producers (Griffin, 2017). The Global South (nations in Africa, Latin America, and Asia) consists primarily of former colonized nations, exploited to extract natural resources and human labor that supported the development of colonial power. In more recent times, the developmental project waged by the Global North claims to increase economic growth and prosperity, raise standards of living, and spread the benefits of consumer culture by integrating developing countries into the global economy (Litonju, 2012).

While the United States continues to have higher per capita emissions than any country, China has replaced the United States as the world's biggest emitter of greenhouse gases, with more than 2.5 times the greenhouse gases emitted by the United States in 2019, and India moved ahead of the European Union to become the worlds' third largest emitter (Larsen et al., 2021; Regan & Dotto, 2021). China's imperial ambitions (and some argue it has become a neocolonial power), through efforts to expand maritime claims, secure global natural resources for economic development, build global infrastructure (Belt and Road Initiative), and enhance its international influence, will further increase global emissions (Blumenthal, 2019; Etzioni, 2020; Yale Environment 360, 2019).

There are important linkages between former colonial policies and practices and those of transnational corporations in more recent times that help explain global divides between major emitting nations and the rest of the world. Williams (2021), writing about a UNESCO World Heritage–protected area of the Sumatran tropical rainforest in Indonesia, found that the wildly profitable Dutch East India Company's success at moving coffee beans from this region to the Western world has largely been taken up by Nestlé, Olam International, and the Louis Dreyfus Company, among several other contemporary international exporters. Williams highlights the ways the Dutch East India Company achieved its success through a forced cash crop planting system that was so severe it resulted in destruction of significant rainforest, mass famine, and starvation. At the end of formal colonial policies, the organizational systems created by the company, such as the public offering of shares, vertical integration of production and processing, and diversified global distribution, became the model for today's multinational corporations. Coffee remains a significant Global South cash crop (with production and distribution managed by Global North multinational corporations for immense profits) that is contributing to the destruction of forests and climate change.

The energy, transportation, agricultural systems, and models of unlimited production and consumption that have characterized Global North industrialization and neoliberal globalization, as well as China's and India's rapid economic development, are increasingly recognized as no longer sustainable. Global North economic policies have reproduced colonial legacies of asymmetrical development within countries and between wealthy and poor nations based on differentiated relationships to global capital (Massey, 1994). The policies and processes of colonialism and developmentalism imposed by wealthy nations on the Global South have resulted in poorer nations and communities disproportionately bearing the brunt of climate effects (Levy & Patz, 2015). Climate denial is core to the refusal by major emitting nations to acknowledge how they contribute the most to global climate change while the consequences of climate change, such as extreme weather events, adverse effects on health and well-being, and impact on livelihoods and environments disproportionately affect lower-income countries and communities (Levy & Patz, 2015).

Petro-nations, like Saudi Arabia, Kuwait, and Russia, have also joined U.S. lobbyists in an effort to water down the IPCC report warning about the dangers of global warming past the limit of 1.5 degrees Celsius (Mann, 2021; Watts & Doherty, 2018). Australia, under Prime Minister Scott Morrison, is also noteworthy for climate denial, rolling back climate policy set by previous administrations and seeking to undermine trust in climate science (Mann, 2021). This small subset of nations has had an outsized role in opposing climate action (Mann, 2021).

The geopolitics of climate denial is a denial of the humanity of vulnerable populations, their cultures, and human rights (e.g., of Indigenous peoples' rights to traditional livelihoods, lands, and security). The climate denial led by Brazilian President Jair Bolsonaro, perhaps the world's most notorious climate denier, effectively mobilized right-wing nationalists, large landowners, agribusiness, and pro-development centrists and resulted in slashing environmental protection budgets, legalizing mining on Indigenous lands, leveling attacks on researchers, and enacting large-scale deforestation of the Amazon (Diele-Viegas & Rocha, 2020; Ferrante & Fearnside, 2019; Goodell, 2021). His assault on Indigenous people and the environment led French President Emmanuel Macron to call Bolsonaro's deforestation policies "ecocide" (Goodell, 2021).

The Philippines, considered the fourth most impacted nation in the world by extreme weather events from 2000 to 2019 (Eckstein et al., 2021), offers a vivid example of Global North-South power dynamics. Climate effects in the Philippines include extreme weather events, threats to food security and biodiversity, rising sea levels, economic loss and increased poverty, and threats to vulnerable groups, such as women and Indigenous people (Climate Change Commission, 2018; NICCDIES, 2021). At the United Nations Climate Change Conference in 2019, the Commission on Human Rights of the Philippines found that climate change constitutes an emergency situation demanding urgent action, that 47 Carbon Major companies played a clear role in anthropogenic climate change and its attendant impacts, and that these contributed to human rights violations (Center for International Law, 2019). None of the 47 Carbon Majors (transnational oil, mining, and cement companies), including Chevron, ExxonMobil, BP, Royal Dutch Shell, and BHP Billiton, which have offices in the Philippines, showed up at the four hearings that were held (Kaminski, 2018; Vidal, 2016).

All maps and visual representations of geographic data tell stories and serve particular purposes and agendas (Segall, 2003). They can reinforce the denial of climate change realities across geographic contexts or vividly represent these realities. The disparities across geographic contexts of the impacts of climate change on nations is represented by maps and geographic data. For example, Our World in Data is one (of many sources) that represent disparities across the world in terms of climate change, emissions per capita, and fossil fuel consumption (by nation, per capita, by energy types, etc.) (Ritchie & Roser, 2020). The Sixth Assessment Report of the IPCC (2021) used graphic representations to convey information about the changing climate and its impacts. *Inside Climate News* (Weisbrod, 2021) published visual representations from the report that summarized key findings, such as changes in surface temperatures, changes in weather and climate extremes by region, and effects on Arctic ice caps, sea level rise, and ocean acidity.

ESRI (the Environmental Systems Research Institute) provides a range of maps on climate migrants and refugees to highlight the impacts climate change is having on vulnerable populations in contexts like Alaska, the world's Island Nations, Darfur, Syria, and Bangladesh (see https://storymaps.esri.com /stories/2017/climate-migrants/). The State of the World Atlas (Smith, 2021) provides a range of cartographic visualizations to examine global patterns, interconnections, and issues, such as international migration trends, the globalization of work, health-related issues (including the COVID-19 pandemic), food and water, and energy resources and consumption. Such graphic representations of the impacts of climate change are powerful reminders of geographic disparities.

SO WHAT BEST EXPLAINS THE RANGE OF CLIMATE BELIEFS IN THE UNITED STATES?

The fossil fuel industry, corporate interests, and well-financed conservative foundations have attempted to discredit climate science and promote fossil fuel and corporate agendas. A climate denial machine has sewn enough doubt or uncertainty about climate science that large percentages of people are cautious (16%), disengaged (7%), doubtful (10%), or dismissive (10%) of the realities of climate change (Goldberg et al., 2020). This climate denial machine has mobilized vast financial resources to influence American politics by ensuring that fossil fuel–friendly information informs public discourse or at least delays climate policy so that industry profits and agendas won't be seriously challenged. The well-established scientific evidence about climate change competes with industry-led and politically motivated messaging, leading to a range of climate beliefs in American society.

Media platforms amplify and spread climate misinformation to foster climate change skepticism and enough distrust or confusion about climate science so that many people think that climate change is part of natural climate cycles and not a serious risk, or that it is a hoax or part of some conspiracy to weaken the United States. Many people are susceptible to climate change misinformation because of the psychological processes of motivated reasoning and identity-protective cognition that nurture online echo chambers that reinforce climate beliefs.

The web of climate denial contexts contributes to increased political sectarianism about issues that are not inherently partisan, like climate change. Climate change continues to be framed as a "debate" to spread misinformation that is often emotionally and politically charged. Often these misperceptions become widespread due to elites spreading them for political advantage (Nyhan, 2014). This impedes the ability of citizens and governments to make decisions based on reliable and valid knowledge, and to make progress on mitigating climate change (Finkel, et al., 2020).

The web of climate denial contexts also shapes and reinforces the dominant stories-we-live-by that are proving to be destructive to our environment and societies: that humans are the center of existence and separate from nature; that the goal of society is economic growth without limits; that humans are motivated by self-interest and most interested in accumulation of material goods; and that nature is primarily a resource to be used for human purposes (Jackson, 2021; Stibbe, 2021). We return to these destructive stories-we-live-by in the final chapter to suggest how to move toward *ecojustice stories-To-live-by*.

An understanding of the contexts of denial and the destructive stories-we-live-by is also essential for critically analyzing climate denial texts. The next three chapters explore how critical literacies+ and eco-civic practices can be used to address the core civic concept of reliability to help confront climate denial.

CHAPTER 4

Denial Texts and Critical Literacies+

In this chapter we explore critical literacies+ that can guide students to examine two types of denial texts: (1) *climate science denial,* which rejects the climate science consensus; and (2) *climate action denial,* which delays action, promotes inaction, or proposes an inadequate response to the climate crisis. The web of denial, as outlined in Chapter 3, shapes the production and distribution of denial texts, which can challenge the capacities of readers to discern reliable, useful information. This is why we ground this chapter (and the next one) with the inquiry question: *Which sources about global warming do we trust and why?* This chapter is a story of how students can answer this question with critical literacies+ that can help them investigate ecologically troubling stories-we-live-by (Stibbe, 2021).

In what follows, we first situate this question as a *civic inquiry,* tied to core issues about what it means to live in a democracy. What to do about climate change, or any issue of public significance, needs to be based on reliable, trustworthy information. Next, we briefly outline an instructional model designed to guide students to investigate and deliberate climate-related texts. Then, we examine an influential *climate science denial* text, highlighting how critical literacies+, namely the FLICC model, can reveal core science denial techniques at play with this text. We also explore the relationship between this denial text and the larger web of denial contexts from Chapter 3. Next, we investigate several *climate action denial* texts. Here, critical literacies+ related to "discourses of delay" (Lamb et al., 2020) and "rhetorical frames" (Supran & Oreskes, 2021) are mobilized. These climate denial texts are tied to stories-we-live-by (Stibbe, 2021) and the chapter concludes with some thoughts about how educators can further support students to confront climate denial and challenge destructive stories-we-live-by.

DETERMINING THE RELIABILITY OF SOURCES

Masyada and Washington (2016) propose a model of civic education that integrates media literacy with conceptual thinking. They draw from core principles of the National Association for Media Literacy Education (NAMLE, 2007) to outline how media literacy education "requires active inquiry and critical thinking about the messages we receive and create" with the goal to develop "informed, reflective and engaged participants essential for a democratic society" (2007). Masyada and Washington then define conceptual as "a focus on the foundational ideas of civics . . . no matter the content that is being studied . . . [ideas that] include such concepts as liberty, equality, security, collaboration, and conflict, among others." They use the C3 Framework from the National Council for the Social Studies (NCSS, 2013), which includes a student-driven and teacher-led inquiry process with four dimensions: ask compelling questions; use disciplinary perspectives of history, geography, economics, or civics; gather and evaluate information sources; and communicate findings and take action. Masyada and Washington then demonstrate their civic education model in action with the concept "civil liberties" and a focused examination of the statement, "The needs of national security might promote restrictions on individual liberty" (p. 91). The authors conclude by stating how their conceptual model, with its grounding in media literacy and commitments to the C3 Framework, reflects an "education for citizenship" perspective, one that centers upon addressing core questions and problems of democratic life.

While Masyada and Washington addressed civil liberties as their concept, we consider *reliability*. To determine if someone is reliable and can be trusted, we typically assess their past performance and our previous experiences with them. We determine whether an information source is reliable by evaluating its provenance (author's background, expertise, experience), purpose (to inform, persuade, etc.), and content (claims, evidence, and corroboration with other sources) (Damico & Baildon, 2015). When we believe the author, creator, or sponsor of a source has done due diligence to ensure the information presented and claims made are accurate, reasonable, and substantiated with quality evidence, we tend to deem the source reliable. Yet, evaluating online sources can be complicated. We are inundated with streams of information, much of it unvetted, and these streams are imbued with the agendas, purposes, and values of authors, agencies, or sponsors. Concepts like reliability are also bound up in broader cultural narratives about what reliability is or means (which we explore in greater depth in Chapter 6).

While "doing source work" has long played a part in social studies inquiry (e.g., Beck et al.,1995; McKeown & Beck, 1994; VanSledright, 2010; Wineburg, 1991, 2001), new forms of information and communication technologies in our Internet age continue to reshape the nature of reading, writing, and communication and necessitate "new literacies" (Lankshear & Knobel, 2003). Evaluating the trustworthiness of sources, assessing how extensively to investigate a source or author, and following links to do additional research are essential online reading behaviors (Liu, 2005). The topics or issues these sources deal with can also be devilishly complex, particularly with "socioscientific issues" like climate change that involve a "social dilemma with conceptual or technological links to science" (Sadler, 2004, p. 513).

Research bears this out. Youth and young adults experience challenges when detecting misleading online information (Wineburg et al., 2016) or the accuracy of online truth claims about controversial public issues (Kahne & Bowyer, 2017). With the complex topic of GMOs (genetically modified organisms), Nicolaidou et al. (2011) found that high school students had difficulty evaluating Internet sources with less clearly defined provenance and needed scaffolding to identify and work through source evaluation criteria, such as genre and funding or sponsors. In some of our own work, we found that high school students were able to move successfully through a set of criteria and questions to evaluate the reliability of a social media web source about health care, yet struggled to gain a larger conceptual understanding of reliability (Damico & Baildon, 2015). Moreover, youth (as well as adults) can exhibit an "optimistic bias" when it comes to their own abilities to skillfully evaluate online information (Flanagin & Metzger, 2010), even when these abilities have been shown to be "bleak" across a range of ages, and up into college (Wineburg et al., 2016). This challenge has intensified in a U.S. social and political context marked by the proliferation of fake news and "alternative facts" (Journell, 2017). Thus, the ways readers of any age evaluate the objectivity of online sources remains of central importance (Cooke, 2017; Marchi, 2012), in particular with divisive socioscientific issues like climate change.

Yet there is some evidence that students can develop these skills. Some studies suggest people can be "inoculated" against misinformation by making explicit who is behind these messages and their motivations, as well as strategies for resisting or refuting them (e.g., Farrell et al., 2019; Lewandosky et al., 2017). Martens and Hobbs (2015) found that students who took part in a school-based media literacy program were able to demonstrate better media knowledge and skills when analyzing advertisements than peers who did not participate in the program. Kahne and Bowyer (2017) highlight that students can more skillfully navigate information and be less prone to biased interpretations after taking part in media literacy training. The opportunity to practice media literacy skills is important (Middaugh, 2017) and teachers need resources to support these efforts. The Civic Online Reasoning Project at Stanford University and the News Literacy Project and its online Checkology platform are indispensable in this regard, offering accessible resources for teachers, administrators, and parents. There is also no shortage of tools, tips, or suggestions for learning how to communicate more effectively about climate change (e.g., Busch & Osborne, 2014). Our goal is to provide teachers and students with the Climate Denial Inquiry Model and tools, such as critical literacies+, for dealing directly with the complex challenges related to discerning climate denial tactics and techniques in texts.

It bears noting that there are different views about what role source reliability should play in social studies—how much instructional time needs to be dedicated to helping students discern whether sources are reliable. In history education, Barton (2005) makes a case to consider reliability in terms of a source's potential as evidence, to pose questions like, "Is there evidence in this source that I can use to help answer my questions?" If the instructional goal is for students to mimic or approximate the work that historians do, evaluating sources for reliability and bias might be misguided. Sometimes the bias contained in a source might be what makes the source reliable because it provides an accurate view of a perspective at a particular time or place. Other than for testimonial accounts, reliability rarely needs to be considered because "it is the very fact of the existence of the source that constitutes historical evidence" (Barton, 2005, p. 747). Thus, if a goal is for students to do the work of historians in the ways Barton points to, going through a process of determining the reliability of sources might seem unnecessary.

We agree that evaluating reliability is often best viewed and understood as part of an analytical process that emphasizes the extraction of useful, relevant evidence to answer questions. Moving outside of historical inquiry (and attempts by middle school or secondary school teachers to help students mirror practices of historians) and into the more general and all-purpose roles and responsibilities of being a citizen, (what we see as central to social studies and education more broadly), we argue it is crucial to have reliable information to reach informed conclusions or to make good decisions about issues such as climate change. As citizens, our first line of defense against making faulty decisions is reliable information. This is highly important because of the misinformation, scams, hoaxes, and unvetted, fake, or doctored information sources that circulate in online spaces. Identifying purposes, weighing bias, and determining the reliability of information is first-order intellectual work to ensure useful, relevant, and sound information is being consulted. Ultimately, if our goal is to best understand an issue, we do not want to waste our time with sources that mislead, misinform, or are outright deceptive.

A Caveat

Whether or not to include and emphasize students' climate beliefs (or other values, perspectives, beliefs, experiences, or knowledge) in classroom inquiries depends on a teacher's goals. In our own teaching, we have found value in including climate belief profiles in an inquiry as well as omitting them. The focus of this chapter is for students to identify and explore climate science denial texts and begin to grapple with broader political and industry-led mechanisms of denial (Freese, 2020; Oreskes & Conway, 2010). While careful consideration of one's climate change beliefs is an important part of this process, it is not essential to make this an explicit part of an inquiry. So, in this chapter, students' climate change beliefs are not emphasized. Their beliefs, however, are central in the next chapter. We do recommend that students complete a climate belief inventory (e.g., Yale Climate Change Communication survey) even if there are no plans to incorporate the survey results in the classroom activity. Completing the survey provides students with additional data about themselves that can support explorations about the relationship between their climate beliefs and their evaluations of climate sources and, by extension, help them articulate their own ecological philosophies or ecosophies. It is also possible that students' climate beliefs might become part of a whole-class discussion.

INSTRUCTIONAL APPROACH

Elsewhere, we have outlined how we have used a particular instructional approach to understand how groups of undergraduate students examine complex, diverse web sources about climate change (Damico & Panos, 2016, 2018; Damico, Panos, & Myers, 2018; Damico, Baildon, & Panos, 2018). We revisit that work here to briefly outline four key features of this approach: (1) juxtapose diverse sources; (2) scaffold multiple traversals of source work; (3) guide critical analysis of denial techniques, discourses, and frames; and (4) facilitate deliberation. We also demonstrate what we see as essential next steps in using critical literacies+ to guide students as they evaluate and deliberate the reliability merits of climate change sources.

Juxtapose Diverse Sources

We want students to work with sources across varied genres, types, forms, and levels of complexity (Damico et al., 2016) and prefer that students work with sources they might encounter on their own online rather than sources located, for example, in restricted library databases. To work with a question like, *Which sources do we trust and why?*, we also want to select sources that oppose or contradict each other—that is, sources that align with the climate science consensus along with sources that reject the consensus. Throughout our teaching we have found this "juxtaposition" strategy—placing texts with conflicting claims, evidence, or arguments side-by-side—to be effective in helping students discern the perspectives, ideas, or ideologies within and across texts. It is also important to locate sources that can engage students in substantive social studies content, defined here as sources situated in the complex web of climate denial as outlined in Chapter 3.

Scaffold Multiple Traversals of Source Work

Reading for reliability can often be more of a recursive than linear process in which readers benefit from "criss-crossing the landscape" of texts to deepen and enrich their meaning-making (Baildon & Damico, 2011a; Spiro et al., 2004). This is why we provide students with multiple opportunities to evaluate the reliability of sources. This includes asking them to explain their thinking for each source several times across an activity by noting and justifying their reliability evaluations (highly reliable, somewhat reliable, somewhat unreliable, unreliable) after each of three traversals (Damico & Panos, 2016).

(1) *Based on a screenshot of each source.* Students do not have access at this point to the full source to scroll through pages, click links, etc. The rationale is that students can benefit from having time to first look carefully at each source, attending to details, piquing interest, or raising questions about the source's url, author or sponsor, organization of tabs, use of images or other figures, etc. Students work individually and write explanations of their reasons for each reliability evaluation. Students can work with partners or in small groups, though we prefer they work independently at this point. After evaluating each information source, they can also rank sources from most to least reliable. This provides them with a baseline of their thinking and a way to track the evolution of their evaluations throughout an activity.

(2) *After viewing each source and answering critical literacy questions.* With this traversal, students have full Internet access to read and evaluate each of the sources. This enables them to investigate various links within each source, open new browser windows to peruse other sources, etc. At this point students also revisit their earlier evaluations based on the screenshots and they document if they want to change their initial evaluations. For example, a student might modify an evaluation from "somewhat unreliable" to "highly reliable" after discerning that there is substantial evidence to support a source's claims and a reputable professional organization is the website's sponsor. As with the screenshot, students can work independently, in pairs, or in small groups, though we have preferred structuring this as independent work because it provides opportunities for them to "be with their own thoughts and ideas" before engaging with others in the upcoming group deliberation.

(3) *After discussing the reliability merits of each source.* With this traversal, students participate in a whole-class discussion about the reliability merits of each source (Which of these sources can we trust?). The instructional goal is for students to have opportunities to persuade or convince each other that their reliability ratings are more plausible than their peers' ratings. After the whole-group discussion, students are prompted to independently document whether or not they want to modify their own evaluations of any of the sources based on the group discussion.

Overall, these multiple traversals highlight distinct opportunities for students to work with sources. Making sense of a screenshot is different from having full access to each source with working hyperlinks, which differs from a whole-group discussion.

Guide Critical Analysis of Denial Texts

The heart of this instructional approach is for students to carefully examine sources. This is where critical literacies come in. With critical literacy questions in mind, students can write responses to a set of questions that emphasize text analysis, such as: *Who created the source? Why was it created? What claims are made? Are claims well supported?* Explain. *Are there biases or points of view? To what extent is this source reliable? (highly reliable, somewhat reliable, somewhat unreliable, or unreliable).* This is also a time for critical literacies+ questions that focus on climate denial techniques, including the FLICC framework (Cook, 2020a), discourses of climate delay (Lamb et al., 2020), frame analysis (Supran & Oreskes, 2021), and stories-we-live-by (Stibbe, 2021). (See Table 2.1 in Chapter 2). These critical literacy+ questions might need to be first taught explicitly as students learn how to work with these tools to examine texts.

Deliberate Ideas

The fourth and final component of this instructional model is for students to participate in a whole-class discussion about the reliability merits of sources. As outlined above, students can draw from evidence in each source to persuade or convince their peers that their reliability ratings are more plausible. The teacher's role during this process depends on how the discussion is unfolding. At times, minimal teacher intervention makes sense as students work skillfully through the deliberation to identify denial techniques and more and less reliable information sources. In other instances, more direct teacher involvement is warranted through, for example, additional questioning or brief presentations of material (concerning what denial techniques are evident in a text). We focus on this fourth component in the next chapter.

CLIMATE SCIENCE DENIAL AND FLICC

Through our teaching and research, we have found that students have had the most difficulty with one climate denial source: the website for the Nongovernmental International Panel on Climate Change (NIPCC). Students have not possessed background knowledge about the NIPCC (e.g., its place in the climate change denial machine), have been uncertain what a nongovernmental status means, and have expressed a need to maintain a "balance" of opposing views. Critical literacies+, especially the FLICC model, can address these issues.

NIPCC Source [http://climatechangereconsidered.org/]

This professional-looking website provides access to a series of reports that have supported the claim that human effects of climate change are "likely to be small relative to natural variability, and whatever small warming is likely to occur will produce benefits as well as costs." When we first used this website with students in 2014, the webpage centered on its report *Climate Change Reconsidered II: Biological Impacts*. The "About" tab on the left stated the purpose of the site was to offer an independent "second opinion" related to the findings of the IPCC (Intergovernmental Panel on Climate Change, a leading climate science authority in the world). NIPCC source was, in fact, created in response to the IPCC with links to the IPCC ("About the IPCC") provided. The webpage also claimed that "procedural problems" with the IPCC have resulted in a "false consensus" about climate change. Since 2014 logos of two organizations, CO2 Science and the Heartland Institute, have appeared at the top right of the home page. When we first taught this activity in 2014, there was a right-side navigation panel that included external links and videos of news clips and featured stories from politically conservative outlets, including Fox News, the Ayn Rand Institute, and Rush Limbaugh.

The website layout, however, changed over time. The right-side navigation and tabs were deleted and replaced with a less text-intensive home page.

Student Challenges with NIPCC Source

When students have worked with just a screenshot of the NIPCC source, a significant majority have deemed it reliable (39% identifying it somewhat reliable, 43% highly reliable). As their reasons, students cite the amount and range of resources and evidence visible in the screenshot and the site's exclusive focus on climate change. Some students also emphasize the NIPCC's international reach, nonprofit status, and the fact that it is not affiliated with the government. However, some students who have rated the site as reliable have expressed concern about potential political bias (especially when the Fox News media link appeared on the bottom right of the screenshot). Students who have found the site unreliable have also noted a lack of data and claims available to view on the screenshot.

After analyzing the full web source with critical literacy questions (Who created the source? Why was it created? What claims are made? Are claims well supported? Explain. Are there biases or points of view? To what extent is this source reliable? (highly reliable, somewhat reliable, somewhat unreliable, or unreliable), reliability ratings have shifted with more students designating it unreliable or somewhat unreliable. However, close to 60% of students have deemed it reliable (including one-third of all students stating that it is highly reliable). The reasons they have provided cluster into three themes. First, some students continued to view the nongovernment status as an indicator of reliability. One student, for example, noted: "Because it's not affiliated w/ the government, they are free to produce their findings w/out fear of government censorship" while another wrote: "Very reliable: non-governmental so they have the earth's best interests at heart and want to educate readers and policymakers on the most recent information available." Second, a number of students have deemed the scientific evidence presented on the site to be compelling enough to justify a positive reliability rating (e.g., "there were reports," "Had lots of sources and years of research"). Third, and most significantly, students have noted the value of NIPCC offering an "opposing" view or "second opinion" on the issue of global climate change, an opinion to counterbalance the scientific consensus.

Web of Denial

While all these students have been asked to answer critical literacy questions about source provenance (Who created the source? Why was it created? Are there biases?), with the NIPCC source, very few have considered the two sponsors, CO2 Science and the Heartland Institute, in their reliability evaluations (e.g., few have clicked on the sponsor links to learn about these organizations). Yet, CO2 Science and the Heartland Institute, as conservative foundations and think tanks, have played a prominent role promoting and disseminating climate denial—namely, by facilitating efforts of "contrarian" scientists to oppose the mainstream or established science (Dunlap & McCright, 2010). As such, both of these organizations are essential cogs in the "climate denial machine" (Dunlap & McCright, 2011).

CO2 Science positions itself as a research and education "center" with its mission statement (as of early 2021) stating:

> The Center for the Study of Carbon Dioxide and Global Change was created to disseminate factual reports and sound commentary on new developments in the world-wide scientific quest to determine the climatic and biological consequences of the ongoing rise in the air's CO2 content. It meets this objective through weekly online publication of its CO2 Science magazine, which contains editorials on topics of current concern and mini-reviews of recently published peer-reviewed scientific journal articles, books, and other educational materials. In this endeavor, the Center attempts to separate reality from rhetoric in the emotionally-charged debate that swirls around the subject of carbon dioxide and global change.

With financial support linked to the fossil fuel industry, CO2 Science and its Center for the Study of Carbon Dioxide and Global Change promote "contrarian" science or climate denial in the web of intersecting contexts of denial (e.g., economic and media contexts).

Also, with economic, political, and cultural-historical contexts of climate denial in mind, the Heartland Institute promotes public policy focusing

on "free-market solutions to social and economic problems" (https://www
.heartland.org/about-us/index.html). Their "About Us" tab indicates that
Heartland is a "national nonprofit research and education organization" and
"one of the world's leading free-market think tanks" as well as a leading
"action tank." Founded in 1984, Heartland gained prominence in the 1990s
with its efforts to discredit the health risks of secondhand smoke and by
challenging bans on smoking. For the past two decades, Heartland has led a
range of climate denial efforts, including the dismissal of the climate scientific
consensus and opposition to any economic proposals or policies that promote a shift away from fossil fuel production and consumption (Klein, 2014;
Oreskes & Conway, 2010).

With this elevated view of the climate denial commitments of these spon-
soring organizations along with keeping the web of denial contexts in mind,
we can mobilize text analysis tools of FLICC to examine the NIPCC source
(*F*ake experts, *L*ogical fallacies, *I*mpossible expectations, *C*herry picking,
*C*onspiracy theories).

FLICC

Perhaps the most obvious denial technique in this source relates to the core
NIPCC claim and "logical fallacy" (the "L" in FLICC) that climate change is
likely caused by natural variability. This example of literal or outright denial
can also be understood as "slothful induction" when relevant evidence is ignored when arriving at a conclusion (Cook, 2020a). This technique is similar
to cherry picking, which emphasizes the selection of some data while ignoring
conflicting evidence. Ultimately, whenever someone or a group considers a
small selection of the available data and draws a conclusion that clashes with
the full body of evidence, it can be understood as cherry picking.

FLICC denial techniques are also at work in the NIPCC's portrait of the
IPCC, which reads:

> The Nongovernmental International Panel on Climate Change (NIPCC)
> was created by a group of scientists concerned about flaws in the organization
> and procedures of another organization, the Intergovernmental Panel on Climate
> Change (IPCC), so it is necessary and appropriate that those flaws be presented
> here.
>
> Though often described by scientists and media as an independent scientific
> organization, the IPCC is in fact an arm of the United Nations. Dr. Steven J.
> Allen reminded us of the true nature of the United Nations in a recent article for
> the Capital Research Center: "The United Nations [is] a famously corrupt body
> in which most votes are controlled by kleptocracies and outright dictatorships.
> Most of the member-states, as they're called, are rated as either "not free"
> or "partly free" by Freedom House, and both Communist China and Putinist
> Russia have veto power. And any settlement of the Global Warming issue by

the UN would entail massive transfers of wealth from the citizens of wealthy countries to the politicians and bureaucrats of the poorer countries. Other than that, one supposes, the IPCC is entirely trustworthy on the issue. (Well, aside from the fact that the IPCC's climate models predicting Global Warming have already failed.)

This depiction of the United Nations (and the IPCC, an "arm" of the UN) as a "famously corrupt body in which most votes are controlled by kleptocracies and outright dictatorships" can be understood as an example of a conspiracy theory, especially the assumption of evil intent. To a certain extent, these types of conspiratorial thinking are inevitable when someone or a group disagrees with the global consensus among climate scientists (Cook, 2020a). A version of an ad hominem argument, a logical fallacy denial technique in which someone assails the motive or character of the person or organization making an argument rather than addressing the substance or evidence of the argument (ad hominem is Latin for "to the person"), is also on display with the NIPCC attack of the UN. Moreover, whether or not "Communist China" and "Putinist Russia" have veto power in the UN is not germane to whether climate change is happening and necessitates swift, extensive action. And stating the IPCC is in any way corrupt requires evidence to support this claim beyond suggesting the IPCC is guilty by association with the UN.

A more charitable interpretation of the NIPCC critique of the IPCC is that it raises potential concerns about how the UN operates. The UN is a large organization with an extensive network of agencies and institutions. Legitimate charges of corruption also mark its past, such as its oil-for-food program in Iraq and senior officials being accused of soliciting bribes (*Corruption at the Heart of the United Nations*, 2005). Yet, the UN has also effectively promoted international human rights, improved health care and alleviated poverty and hunger in many countries, and influenced international relations and international law by brokering hundreds of multilateral treaties for peace, security, and the protection of rights and the environment (Chesterman et al., 2019; Gordenker, 2017).

In sum, using critical literacies+ of FLICC can help readers confront denial in this type of climate science denial text. Given that information about the relationships among the NIPCC, its sponsors (CO2 Science, the Heartland Institute, and the Science and Environmental Policy Project) and fossil fuel companies is not readily apparent on the NIPCC website (the site maintains the NIPCC represents the work of independent, nongovernmental scientists and scholars), it is also important for students to evaluate the sponsors of this website (CO2 Science and the Heartland Institute). This requires leaving the website to do what Stanford University's Civic Online Reasoning project calls "lateral reading" to see what other digital sources say about a source, who's behind the information, and what their agendas might be (Wineburg & McGrew, 2019).

CLIMATE ACTION DENIAL: DELAY DISCOURSES AND FRAMES

Another type of climate denial text acknowledges the global warming scientific consensus yet promotes inaction or an inadequate response to the crisis. One prominent climate action denial text that we have used in our teaching since 2014 comes from the oil and gas multinational giant BP (formerly British Petroleum and rebranded as "Beyond Petroleum" in the early 2000s).

BP and the CDIM

BP is among the leading transnational oil and gas companies in the world. Back in 2014, when we began teaching about its response to climate change, BP's webpages included a concise summary of the climate science beneath the image of two workers at an oil refinery. BP also identified climate change as an "important long-term issue that justifies global action" and that "meeting the energy challenge" was a top priority.

In recent years, however, BP has attempted to position itself as more forward-looking and strongly committed to addressing the escalating climate threat. In February 2020 BP announced a goal to become a net-zero company by 2050 by eliminating or offsetting all planet-warming emissions by this time. BP also committed to cut oil and gas extraction 40% by 2030. This overhaul of the company's mission and business strategy is highlighted in BP's new "sustainability frame" with its three focus areas: "getting to net zero," "improving people's lives," and "caring for our planet" (https://www.bp.com /en/global/corporate/sustainability.html). Of note, the website has shifted its emphasis from "Climate change and the energy transition" to "sustainability" (though it is possible the emphasis will continue to evolve and change).

With its revised website, BP also goes to some length to proclaim plans and outline aims to transform its structure and operations as BP joins a growing list of corporations and governments with commitments to achieve net-zero greenhouse gas emissions by 2050.

When students have responded to critical literacy questions about the authorship and purposes, claims, evidence, bias, and reliability of the BP source (e.g., Who created the source? Why was it created? What claims are made? Are claims well supported? Explain. Are there biases or points of view? To what extent is this source reliable?), many have expressed concerns about its reliability, given its status as a transnational oil company and the 2010 BP Deepwater Horizon oil spill in the Gulf of Mexico (the largest marine oil spill in history; most students have remembered this event). This has led some students to argue that this BP climate-related site is "just a publicity stunt" or a "textbook case of greenwashing."

Yet, a surprising number of students have found this source at least somewhat reliable (and more than expected have deemed it highly reliable), noting that it seems to present accurate information and, when it comes to deciding what might be done to address climate change, BP's perspective is important because of its significance in our energy economy. Moreover, this source points to the ways BP is addressing the climate change threat, demonstrating, for example, its production shift to "lower carbon fuels" and more sustainable solutions. Students have also identified being surprised by BP's emphasis on climate change, finding it reliable, in one student's words, "because my gut would have been that an oil company would refute the existence of global warming because of their own contribution [to climate change–related] human activities."

Of the dozens of students we have worked with, almost all have not identified BP as a key cog in the "climate denial machine" (Dunlap & McCright, 2011), situating this transnational oil company within wider contexts of denial as described in Chapter 3—that is, oil companies have persisted with their extractive model of petro-capitalism while leading efforts to sow doubt and confusion about climate science through highly funded media and political campaigns (Freese, 2020; Oreskes & Conway, 2010). This is why climate denial questions and concepts about specific denial techniques, discourses, and frames are useful. It is true that since 2014, when we began working with BP's webpages about its climate change stance, BP has not engaged in literal or outright climate denial. The summary of climate science, for example, has been straightforward with no major distortions. The mode or "state of denial" (Cohen, 2001) employed has been more "interpretive," with climate change facts distorted or "spun" into unwarranted interpretations, and "implicatory," marked by a refusal to both acknowledge and sufficiently act upon the implications of the science.

In this way, an emphasis on discourses of delay (Lamb et al., 2020) within a "fossil fuel savior frame" (Supran & Oreskes, 2021) can help students more fully grasp the workings of denial in these webpages. Within the fossil fuel savior frame, BP presents itself as "reimagining energy" to improve people's lives and care for our planet, while continuing to "create value" for stakeholders (society, employees, and shareholders). The website offers an example of fossil fuel solutionism that highlights technological optimism (touting BP innovation and technology) and the continued value of fossil fuels for the economy, society, and the future.

Students can also explore the discourses of "redirecting responsibility" and "pushing non-transformative solutions" along with the ways that fossil fuels are cast as essential to meet longer-term energy needs. Oil companies have also diverted attention from fossil fuel regulations and instead have championed efforts for current and future (and more uncertain) technologies (Lamb et al., 2020). BP's framing of climate change as a "long-term issue" also evinces a discourse of delay.

BP has also played an outsized role in cementing an "individualized responsibility" frame in the wider culture, which has had remarkable staying power. In 2004, BP, with the aid of public relations professionals Ogilvy and

Mather (Solnit, 2021), launched the highly influential "carbon footprint" campaign, which promulgated the view that the responsibility for reducing carbon emissions resided with individuals rather than institutions or corporations. This, in effect, framed the economics of climate change as a demand-side problem (people's energy needs) rather than a supply-side problem (fossil fuel energy that companies are providing needs to change).

A closer look at BP's 2050 net-zero pledge also reveals more "interpretive denial" at work. For starters, the net-zero target relies on the capture or cancellation of emissions through "unproven technologies and reforestation at a questionable scale" so attaining net zero does not mean greenhouse gas emissions will end (Kusnetz, 2020, para. 7). Along with other oil companies, BP also indicated its intention to reduce the "carbon intensity" (pollution per unit) of its products, but total emissions might increase if it sells more solar or biofuels to counteract its gas and oil sales (Kusnetz, 2020). BP's commitment to cut oil and gas extraction 40% by 2030 also excludes its major share of the Russian oil company, Rosneft, which accounted in 2019 for roughly 30 percent of the carbon pollution related to BP's extraction investments (Tong, 2020). Moreover, BP has remained an active member of lobbying groups opposed to climate actions in the United States and Australia (Tong, 2020). According to an investigative report, these groups are part of a "sprawling network of state and regional trade associations that have, in at least one case, boasted about quashing the very carbon-reduction policies the oil giants publicly claim to support" (Boren et al., 2020, para. 4).

BP also provides a case of using the technique of paltering, a form of greenwashing in which companies make claims that may be true yet are misleading (Cook, 2020a). BP appears to embrace global climate action and claims to be shifting production to more sustainable solutions while failing to mention, for example, the actions of Rosneft, the Russian oil company that BP has significant investments in (a 19.75% share, according to BP's website; Tong, 2020). Rosneft ranked first in revenue for all Russian companies in 2020 and has increased its overall oil and gas production with new projects not only in Russia but in Egypt, Brazil, Iraq, Vietnam, and Mozambique as well (Cooper, 2021). BP, along with other major fossil fuel corporations, has also used misleading accounting strategies to "creatively reclassify, bury and entirely exclude their total emissions" (Eisenfeld et al., 2022, p. 7). This includes their attempts to exclude "scope 3" emissions (a range of indirect emissions in their supply chains), which can comprise more than 75% of total emissions related to oil and gas production.

BP is also among the world's five largest oil companies that have collectively spent more than $1 billion since the 2015 Paris Climate Accords to rebrand themselves as "green" while continuing efforts to secure new oil sources and impede governmental climate regulations (InfluenceMap, 2019). Given the sophistication of BP's efforts to situate itself as a more progressive oil and gas company, students need opportunities to identify denial techniques, discourses, and frames as outlined above. Now let's consider examples from former U.S. presidents.

Trump Response to Question About Climate Change

This is a response from then-President Donald Trump during a press conference at the end of the 2019 G-7 Summit. The question posed to Trump was: *"Mr. President, there was a significant talk at the summit about climate change. I know in the past you've harbored some skepticism of the science on climate change. What do you think the world should be doing about climate change? And do you still harbor that skepticism?"*

I feel that the United States has tremendous wealth. The wealth is under its feet. I've made that wealth come alive. We will soon be one of the—we will soon be exporting. In fact, we're actually doing it now—exporting. But we are now the number one energy producer in the world. And soon, it will be by far, with a couple of pipelines that have not been able to get approved for many, many years. It'll have a huge impact. I was able to get ANWR [Arctic National Wildlife Refuge] in Alaska. It could be the largest site in the world for oil and gas. I was able to get ANWR approved. Ronald Reagan wasn't able to do it. Nobody was able to do it. They've been trying to do it since before Ronald Reagan. I got it approved. We're the number one energy producer in the world. Soon it will be, by far, the number one. It's tremendous wealth. And LNG is being sought after all over Europe and all over the world, and we have more of it than anybody else.

And I'm not going to lose that wealth. I'm not going to lose it on dreams, on windmills—which, frankly, aren't working too well. I'm not going to lose it. So, Josh, in a nutshell, I want the cleanest water on Earth. I want the cleanest air on Earth. And that's what we're doing. And I'm an environmentalist. A lot of people don't understand that. I have done more environmental impact statements, probably, than anybody that's—I guess I can say definitely, because I have done many, many, many of them. More than anybody that's ever been President or Vice President or anything even close to President. And I think I know more about the environment than most people. I want clean air. I want clean water. I want a wealthy country. I want a spectacular country with jobs, with pensions, with so many things. And that's what we're getting. So I want to be very careful. At the same time -

At the same time, it's very important to me—very important to me—we have to maintain this incredible—this incredible place that we've all built. We've become a much richer country. And that's a good thing, not a bad thing—because that great wealth allows us to take care of people. We can take care of people that we couldn't have taken care of in the past because of the great wealth. We can't let that wealth be taken away. Clean air, clean water. Thank you very much everybody. I appreciate it. Thank you. Thank you very much. (https://www.youtube.com/watch?v=pl1Rnz4zNkg)

Of note, this is not a climate science denial text. Trump does not claim that climate change is not occurring or is a "hoax"; nor does he employ other techniques that merit using FLICC tools. When asked about climate change, climate science, and what the world should be doing about climate change, he talks about U.S. energy production and wealth. Trump is using techniques of preemptive framing (by emphasizing energy production and his political ability to get pipelines approved and open ANWR to drilling), diversion (diverting attention away from climate change), and deflection (changing direction of the focus to his political skill while claiming to be an environmentalist; Lakoff, 2017).

In terms of discourses of delay, Trump is very much wedded to fossil fuel solutionism (Supran & Oreskes, 2021), as he extols the benefits, if not the virtues, of domestic energy production, framing it as the fundamental source of wealth in the country and something that he is not willing to "lose." Maintenance of the existing fossil fuel–based economic model is the priority; nothing will jeopardize this. He also gestures toward the discourses of "energy poverty" (moving away from fossil fuels would jeopardize prosperity and lead to greater poverty) and "emphasizing the downsides" (making substantive changes will be too disruptive and will burden society) with his dismissive comment about transitioning to renewable energy sources like windmills.

Obama and Offshore Drilling

On the last day of March 2010, then-President Obama made a speech to announce a proposal to open large regions of water along the Atlantic coastline, the Gulf of Mexico, and coast of Alaska to the drilling of oil and natural gas. He stated:

> We need to make continued investments in clean coal technologies and advanced biofuels. A few weeks ago, I announced loan guarantees to break ground on America's first new nuclear facility in three decades, a project that will create thousands of jobs. And in the short term, as we transition to cleaner energy sources, we've still got to make some tough decisions about opening new offshore areas for oil and gas development in ways that protect communities and protect coastlines.
>
> This is not a decision that I've made lightly. It's one that Ken and I—as well as Carol Browner, my energy advisor, and others in my administration—looked at closely for more than a year. But the bottom line is this: Given our energy needs, in order to sustain economic growth and produce jobs, and keep our businesses competitive, we are going to need to harness traditional sources of fuel even as we ramp up production of new sources of renewable, homegrown energy.
>
> So today we're announcing the expansion of offshore oil and gas exploration, but in ways that balance the need to harness domestic energy resources and

the need to protect America's natural resources. Under the leadership of Secretary Salazar, we'll employ new technologies that reduce the impact of oil exploration. We'll protect areas that are vital to tourism, the environment, and our national security. And we'll be guided not by political ideology, but by scientific evidence.

That's why my administration will consider potential areas for development in the mid and south Atlantic and the Gulf of Mexico, while studying and protecting sensitive areas in the Arctic. That's why we'll continue to support development of leased areas off the North Slope of Alaska, while protecting Alaska's Bristol Bay. (https://www.nytimes.com/2010/04/01/science/earth/01energy-text.html)

As with the excerpt from Trump, this statement from Obama is not an example of science denial; it is action denial. In Obama's speech the frame of technological optimism is employed. Obama makes the case that new technologies will reduce the impact of oil exploration in the mid and south Atlantic, Gulf of Mexico, and the Arctic and help the United States eventually transition to cleaner energy sources. In terms of a technological shell-game frame, Obama also highlights clean coal technologies, advanced biofuels, and nuclear power as part of a "strategic ambiguity" (Schneider et al., 2016) in eventually making this transition. Obama's emphasis on opening large coastal regions along the Atlantic, Gulf of Mexico, and the coast of Alaska to new offshore drilling to transition to cleaner energy while claiming to protect communities and coastlines also suggests technological shell game.

Obama also uses a denial frame of fossil fuel solutionism to argue that his energy policy creates jobs, sustains economic growth, and keeps businesses competitive. His speech suggests a type of greenwashing as a political and public relations strategy to portray an ecologically responsible image for his administration and policies. He uses the rhetoric of balance—to balance "traditional sources of fuel" with renewable energy sources, and economic growth based on fossil fuels with protecting the environment. He notes the need to protect coastal areas, but for tourism, and highlights how new oil and gas sources will enhance national security while claiming this balancing act is not guided by political ideology but by scientific evidence.

We selected these two sources from former U.S. presidents not to suggest that there were no substantive differences between Obama and Trump when it came to their climate policies. It was the Obama administration that issued the Clean Power Plan, as a presidential executive order, to provide guidelines and support to states for reducing carbon emissions as a way to help the United States abide by the Paris Agreement. The Trump administration withdrew the United States from the Paris Agreement, repealed the Clean Power Plan, lifted several bans on oil and gas exploration, and took dozens of deregulatory actions deemed to have weakened environmental protections (Gross, 2020). What the juxtaposition of these two sources does highlight is how prevalent and even normalizing climate action denial can be.

CLIMATE DENIAL TEXTS AND STORIES-WE-LIVE-BY

The four denial texts in this chapter—NIPCC website, BP webpages dedicated to climate change or sustainability, Donald Trump press conference excerpt, and Barack Obama speech excerpt—reflect and reinforce several core stories-we-live-by. The story that humans are the center of existence and nature is a resource to be used and exploited undergirds each of these texts. While some of these texts express concern for the environment, the energy needs of humans remain the priority. A companion story-we-live-by—that a primary goal of society is economic and technological growth—also permeates these denial texts with an unwavering commitment to fossil-fuel solutionism.

CONCLUSION

The Climate Denial Inquiry Model offers critical literacy+ tools for teachers and students to identify climate denial techniques at work in texts. Applying these techniques is essential in helping students grapple with a core civic question of our time, *Which sources do we trust and why?*, and helps clarify how denial texts link to or index different stories-we-live-by.

Doing this work in classrooms, however, can be challenging. To engage thoughtfully with the four denial texts above demands a certain level of background knowledge and familiarity with fossil fuel propaganda efforts and perhaps at least a rudimentary understanding of the climate change denial machine. Teachers and students can build and fortify their knowledge with resources like Inside Climate News and DeSmog that offer ongoing reports and investigative journalism that confront denial. Texts or information sources that directly confront students' inaccurate understandings of scientific phenomena are sometimes called "refutational readings" (Sinatra & Broughton, 2011).

The next chapter continues an exploration of how to address the question, *Which sources do we trust and why?*, as we pivot to eco-civic practices of deliberation and reflexivity, additional essential tools for confronting climate denial.

CHAPTER 5

Eco-Civic Practices of Deliberation and Reflexivity

"WE REALLY SHOULD HAVE THOUGHT ABOUT THIS"

It is mid-February as a group of 25 undergraduate social studies education students are about to deliberate the reliability merits of the NIPCC source. This group of students (along with many of the 150+ students we have worked with on this classroom activity) mostly viewed both the NIPCC and the IPCC sources as reliable because many students perceived a need to include "second opinions" or have "both sides" represented about climate change. As this group of students is set to begin the whole-class deliberation (the fourth and final phase of an instructional model outlined in Chapter 4), 80% have deemed the NIPCC source at least somewhat reliable (with more than 40% deeming it highly reliable). Again, a primary goal of this whole-class discussion is for students to have opportunities to persuade each other their reliability ratings are better or more plausible than their peers' ratings. The teacher's role during this process depends on how the discussion is unfolding.

As the instructor, I [James] am using a discussion format (Damico & Panos, 2016) where I have staged the classroom in four quadrants: highly reliable, somewhat reliable, somewhat unreliable, and unreliable. To discuss a source, I ask students to move to the quadrant that aligns with their reliability ratings. This is after the students have completed two traversals with the sources: their analysis of the screen shot of the source and their analysis after examining the full source more closely based on a set of questions to promote critical reading (Who created the source? Why was it created? What claims are made? Are claims well supported? Explain. Are there biases or points of view? To what extent is this source reliable? See Chapter 4 for more details about this instructional approach).

I have also reminded them that they can move to another quadrant at any point in the discussion if a classmate from another quadrant offers a persuasive, convincing argument. All they need to do is walk across the room and join another quadrant to signal a change in their own reliability evaluation. We now begin working with the NIPCC source and I ask the students to head to the quadrant of the room that aligns with their evaluations for this source. Students make their way across the room and almost all students have placed themselves in either the "somewhat reliable" or "highly reliable" quadrants. Only two students park themselves in the "unreliable" corner of the room.

James: Ok, let's start with the highly reliable quadrant. What are your reasons for your evaluations?

Bethany: The IPCC is headed by the UN (United Nations) which is politics and interest groups so the reason why we are "strongly agree" [view the NIPCC as highly reliable] is because we are just looking at the science. We are not looking at the politics.

Daniel: Wait, but the Heartland Institute who is funding the NIPCC is a conservative think tank from Illinois. Just FYI.

Students: [some laughter]

James: [to Daniel] Do you want to say more about that?

Daniel: The Heartland Institute?

James: Yes, or . . .

Daniel: Looking at the Heartland Institute and other organizations they fund . . . they are typically very conservative maybe even leaning towards Christian conservative values. Not necessarily perhaps, but to me trying to say that that [IPCC] is political and the other [NIPCC] isn't is just categorically false because yeh, you have certain interests within the UN that might speak louder than others but they [NIPCC] call themselves a conservative institute so putting money towards that is saying something significant.

The students standing in the "highly reliable" quadrant are now taking out their phones, looking up information about the Heartland Institute. All but one student then move across the room and join Daniel in the "unreliable" quadrant. As they re-situate themselves, I lead a brief exchange about the Heartland Institute as students cite what they just learned from quick research on their phones. They identify, for example, the Heartland Institute's long-standing role and impact in disseminating climate denial as well as its lobbying efforts to oppose smoking bans. I hear an exchange between two students. One student states, "Wow! I didn't even think about this. It's pretty basic." The other responds, "Yeh, we really should have thought about this."

In this whole-class discussion students deliberated whether or not an information source was reliable. One student, Daniel, was able to convince his classmates that a climate science denial source was not as reliable as they thought. Daniel challenged the claim of his classmate, Bethany, who contended that the IPCC was about "politics" and "interest groups" while the NIPCC source focused on "the science." In a brief dialogue across difference with his peer, Daniel impugned the credibility of the NIPCC website with a straightforward analysis of the source's sponsor, the Heartland Institute. His classmates then independently verified the perspectives and past efforts of the Heartland Institute (conducted web searches about Heartland on their phones), before they changed their reliability designations. This helped them begin to historically situate the Heartland Institute, and the NIPCC source, as part of a larger and longer trajectory of science denial (i.e., how the Heartland Institute challenged evidence about adverse effects of smoking). Overall, students arrived at this insight about the NIPCC source based on just a brief exchange between two students that a teacher (James) facilitated.

This vignette provides an example of how a group deliberation process can help students work across different understandings and interpretations of a climate denial text. Given the emphasis on climate change and source reliability (a core civics concept as outlined in Chapter 4), this was an eco-civic deliberation with students coming to a near unanimous agreement about the NIPCC source. This outcome was also possible because this group of students valued the practice of careful examination of claims and evidence. During this semester the students were all enrolled in the same cluster of three social studies education courses, and this core reading practice was foundational to these courses. This was a group that took pride in being critical readers and in learning with and from each other as a small community of future social studies educators. So, it is not surprising that almost all of Daniel's classmates were willing to change their minds based on the evidence he presented about the NIPCC source.

While Chapter 4 highlighted the ways students can use critical literacies+ to investigate climate denial texts and discern the reliability of sources, this chapter explores the eco-civic practices of deliberation and reflexivity as tools educators can mobilize to help students respond to climate denial texts. As the vignette above demonstrates, teachers can design opportunities for students to collaborate and talk with each other as they jointly investigate texts across their differences to bolster their knowledge about climate denial. The rest of this chapter considers ways teachers can intentionally guide students to be reflexive in their deliberations with any type of climate denial text. Given that the ways we read are shaped by or "motivated" by different dimensions about ourselves, or the multiple identities that influence how we see the world, it is important to first outline work related to motivated reasoning.

MOTIVATED REASONING AND REFLEXIVITY

In *The Lies That Bind: Rethinking Identity*, Kwame Anthony Appiah reminds us that identities matter to us because they help us see how to "fit into the social world"; identities come with ideas and labels, shape our thoughts about how to behave, and affect how other people treat us (2018, p. 12). Classifying or categorizing identities is also complicated. Appiah considers five identities (creed, country, color, class, culture) in his classification amidst a number of other possibilities, which inevitably overlap and intersect (e.g., age, sexual orientation, political ideology, dis/ability). Much research has shown that the ways we make meaning with texts are "motivated" or shaped by an emotionally driven process to seek out and interpret information that confirms one's existing views or beliefs. Given the influence of motivated reasoning, confirmation and disconfirmation bias, and identity-protective cognition, we are more likely to dismiss or reject new information that challenges what we already believe. These motivations are even more pronounced when we perceive a threat to our sense of identity or the status of our identity group or "tribe." This helps explain why online readers are likely to consume perspectives that align with their own (Jamieson, 2008; Kahne & Bowyer, 2017; Prior, 2013). Political affiliations, for example, can lead to different beliefs about fundamental facts around divisive topics, such as the Iraq War (Kull et al., 2003), income inequality (Bartels, 2009), climate change (Dunlap & McCright, 2011), results of a U.S. presidential election, and the COVID-19 coronavirus.

We have long conceived of critical literacies as two sets of core practices—text analysis and reader reflexivity—as flipsides of the same coin (Baildon & Damico, 2011a). While text analysis emphasizes careful scrutiny of sources (determining purposes, evaluating claims and evidence, etc.), critical self-reflection or reflexivity turns that analytical lens inward, asking readers to investigate what they bring to texts, including their beliefs, values, knowledge, identities or social locations (e.g., gender, race, nationality, religion, social class, among others) because "where we read from matters" (Damico, Baildon, Exter & Guo, 2009). And because reader reflexivity tends to receive less attention than text analysis in the curriculum materials and instructional routines across the many classrooms where we have worked or visited, integrating reflexivity into classroom practices can help students better understand and address how motivated reasoning shapes their meaning-making.

In this sense, reflexivity is both a literacy practice and a civic practice. Reflexivity turns the analytical lens inward, asking individual readers to grapple with what they bring to texts, including ideas and emotions. Reflexivity also includes being aware of the effects our ideas, language, actions, and emotions have on our social interactions and relationships, including how we might use language effectively to influence others in public spaces and make collective decisions (Allen, 2004; Tully, in press). It is care and concern for the consequences of one's actions on others (Maturana, 2008), an understanding that what we do affects others and what others do affects us. Reflexivity involves recognizing our perspectives, positionalities, and agency and includes capacities to analyze power relations, structures/systems, and knowledge that disempowers others (Andreotti, 2006; Hauerwas et al., 2021). Becoming more reflexive, learners can enrich and deepen discussions across differences.

AGONISTIC ENCOUNTERS AND
TRANSFORMATIVE INTERROGATION

Contestation and compromise are central to living in a democracy and addressing issues of public concern. These social processes are also central to knowledge-building. Science (as with all academic disciplines) is a social activity that involves deliberation over claims and evidence, the rigorous critical scrutiny and contestation of arguments, and revision of arguments as a matter of compromise if errors are detected or better arguments proven to be true (Oreskes, 2019). Guiding students to be more reflexive (and aware of their motivated reasoning) similarly requires the creation of learning experiences in which they encounter others (e.g., in classrooms or others' ideas in the form of texts, counternarratives, etc.) who challenge their reasoning, offer new insights, or provide "better" ways to understand texts and issues.

Drawing on work by political theorist Chantal Mouffe, we use the term *agonistic encounters* to describe how teachers can create opportunities for students to dialogue across differences. Agonism, or agonistic pluralism, is a social and political theory that emphasizes the role that conflict plays in democratic processes as competing ideas and perspectives vie against each other. While different or opposing beliefs and ideas may be irreconcilable at times, people with different views are considered to be adversaries whose existence is legitimate rather than enemies to be vanquished (Mouffe, 2013).

In thinking about designing agonistic encounters, teachers can keep three goals in mind (Sant et al., 2021):

(1) normalize disagreement to help students understand that disagreement is democratic and central to deliberation and is not about conflict for its own sake;

(2) create opportunities for the expression of political emotions. Because political topics or projects often spring from within ourselves, "bound up with basic human drives and desires" including the "need for collective identifications" (Ruitenberg, 2010, p. 46), teachers can encourage students to engage passionately in a deliberation rather than require them to censure their emotions; and

(3) cultivate opportunities for new identities, positions, or levels of conscious awareness to emerge, which can happen when students are asked to engage substantively with new or different discourses, storylines, or perspectives.

In agonistic encounters, students come together to evaluate texts, offer different interpretations, and engage with varied perspectives to make decisions about reliable information that will enable them to understand climate change or determine climate action. In the process students ideally interrogate their own reasoning, views, and positions to disrupt the "internal narratives" that often justify particular perspectives and positions that perpetuate social problems (Teo, 2017). Deliberation across difference requires students to critically interrogate their ideas and positions yet remain engaged with each other, committed to existing together-in-plurality, with the recognition that difference holds potential for the revision and improvement of ideas and practices (Arendt, 1958). The philosopher of science Helen Longino (2020) refers to these moves of criticism, revision, and potential acceptance of new or better ideas to build knowledge as occurring through a process of "transformative interrogation" (p. 212). As teachers, we hope agonistic encounters facilitate transformative interrogation as students with varied beliefs and views develop shared understandings about denial texts and how they operate.

Given the ways that readers are "motivated" to interpret and understand texts in particular ways, how might educators explore the potential of agonistic encounters and transformative interrogation in which students can access and mobilize their varied beliefs, values, experiences, or identities as they engage with each other? Next, we use an example of an intentionally designed agonistic encounter between two undergraduate students who discuss a climate denial source. Unlike the whole-class discussion that opened this chapter, this example exhibits no radical shift in thinking or gestures toward transformation. Rather, it serves as a springboard to think through ways teachers can guide students to engage across differences through deliberation and reflexivity.

DESIGNING AGONISTIC ENCOUNTERS WITH CLIMATE DENIAL TEXTS

Deliberation Across Climate Beliefs and Academic Disciplines

Two ways to promote potential agonistic encounters is for students to work across their varied climate change beliefs and across different academic subject areas, disciplines, or interests. After students completed the four phases of the learning activity, as outlined in Chapter 4, we created a follow-up learning opportunity for students to work in pairs and evaluate additional climate change sources. In one pair, John was a science education major who said that he had "a lot of background knowledge about climate change" due to high school debate experiences and his science focus in college. His Yale Climate Change Communication survey profile was "alarmed." Todd was a social studies education major who planned to teach history in high school. He stated that he was not interested in climate change because it was "not a pressing issue." While working on the initial activity, he wrote, "Let's take out ISIS first and move on to Iran and North Korea next. After that we can balance the budget and maybe get to climate change." His climate change profile was "dismissive." Figure 5.1 highlights how these two students with distinct academic or disciplinary emphases occupied the opposing ends of the climate belief spectrum based on their results with the Yale Climate Change

Figure 5.1. Two Students with Opposing Climate Change Beliefs

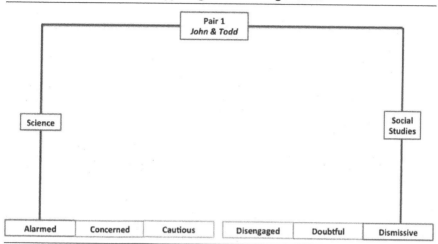

Communication survey. When they began their work, the students were aware of each other's climate belief differences and content area foci.

To further outline differences between these students, Table 5.1 presents how John and Todd evaluated the NIPCC and IPCC sources in the initial activity, which highlights, for example, Todd's concerns about the credibility of government-sponsored websites.

It bears noting that surveys to identify climate change belief profiles are useful but limited. Given Todd's dismissive orientation, it might be easy to label him a staunch climate denier. Yet, his stance seemed less driven by rigid ideological commitments than by practical considerations. At one point in his deliberation with John, Todd said, "A lot of climate change dismissives would not have a problem with alternative fuel. It's not like every climate change dismissive is a big oil fat cat who drives their Hummer down the road." He added, "If alternative energy was of equal price and of equal availability as fossil fuels then climate change no longer becomes the actual issue. Then people wouldn't care, and we could end up saving the climate that way."

John and Todd evaluated two additional sources as part of the follow-up activity. The first text was a 60-second climate science denial video about the benefits of carbon dioxide; the second text was a policy advocacy video to combat climate change titled "What is fossil fuel divestment and why does it matter?" from *The Guardian*, a well-known British newspaper with a global reach that touts itself as "the world's leading liberal voice." Elsewhere we have described how these students worked with these sources, enacting critical literacies through an agonistic process (Damico, Panos, & Baildon, 2018). We expand upon that work here. Like texts, which are motivated

Table 5.1. Student evaluations of sources

	John (Science, alarmed)	Todd (Social Studies, dismissive)
IPCC source		
Is this source reliable?	Seems pretty reliable. Again, all the scientists and the Nobel prize on the home page have won me over.	It's a government website; need I say more. As credible as any government website can be. Take that how you will.
Are there any biases? Explain.	It's hard to find bias in a scientific report since researchers can cherry-pick data from wherever they want to support their opinions. It could be biased in some ways, but I'm inclined to believe the thousands of scientists that support this data.	Climate change is real. Deal with it.
NIPCC source		
Is this source reliable?	I don't find it very reliable, but it sure does look the part. All the pictures and videos make it look like a lot of other sciency websites that I've seen.	As reliable as anything with an agenda can be.
Are there any biases? Explain.	All the attacks against the IPCC without actually justifying their claims seem to me to be pretty biased in the favor of climate change deniers. Merely on the front page of the website, though, it is hard to discern any explicit biases.	The belief that because this is not a governmental source it is obviously more reliable.

(have purposes and agendas), readers are motivated; they come to texts with their own beliefs, values, knowledge, experiences, and identities.

Climate Denial Text

Below (Table 5.2) is a full transcript of the denial video titled *Global Warming: Glaciers*, with a brief description of the images used to accompany the spoken text and reinforce characteristics of science denial across the 60-second video (https://www.youtube.com/watch?v=Wq_Bj-av3g0).

   This YouTube video was produced by the Competitive Enterprise Institute (CEI), a politically conservative think tank founded in 1984 that identifies itself as a nonprofit public policy organization aimed at "advancing the principles of limited government, free enterprise, and individual liberty." In terms of policy issues, CEI has focused on the environment and energy

Table 5.2. Global Warming: Glaciers *Video Transcript with Image Descriptions*

Time	Spoken text	Images
00–:03	You've seen those headlines about global warming.	*Time* magazine cover and *Washington Post* March 3, 2006, front page headline, "Ice Sheet Is Melting Rapidly"
04–:08	The glaciers are melting. We're doomed.	Image of glacier collapsing
09–:12	That's what several studies supposedly found.	Image of *USA Today* article darts across screen. It is titled, "Greenland Glacier Runoff Doubles over Past Decade." We an see the date on the newspaper page. It is February 17, 2006. Then transitions to image of an article titled, "Glacier Melt Could Signal Faster Rise in Ocean Levels." The video creator added the following attribution at the bottom of the screen: the *Washington Post*, front page, February 17, 2006
13–:16	But other scientific studies found exactly the opposite.	Original image of glacier collapsing reverses and returns to precollapsed state
17–:19	Greenland's glaciers are growing, not melting.	The cover of the journal *Science* is displayed and transitions into an article titled, "Recent Ice-Sheet Growth in the Interior of Greenland." A date is not discernible.
20–:23	The Antarctic ice sheet is getting thicker not thinner.	Again, there is a quick glimpse of the cover from the journal *Science*, which opens to a title of an article, "Snowfall-Driven Growth in East Antarctic Ice Sheet Mitigates Recent Sea-Level Rise." The word "Growth" is lightly illuminated. Again, no date is discernible.
24–:26	Did you see any big headlines about that?	A blank page with "The Washington Post" at the top and a big question mark in red fills the page.
27–:28	Why are they trying to scare us?	
29–:34	Global warming alarmists claim the glaciers are melting because of carbon dioxide from the fuels we use.	Image is mostly snow in the foreground with mountains in background. One-by-one the following names appear on the screen: Sierra Club, NRDC, Friends of the Earth, Greenpeace.
35–:37	"Let's force people to cut back," they say.	Short video clip of someone riding a bike in wintry mix of snow, ice, or rain next to a busy street.

(continued)

Table 5.2. Global Warming: Glaciers *Video Transcript with Image Descriptions (Continued)*

Time	Spoken text	Images
38–:44	But we depend on those fuels to grow our food, move our children, light up our lives	Image of refinery plant. It is early morning with yellow sun appearing toward the bottom left. Quickly pans to a tractor in a field then several young girls getting into seemingly a minivan. Then pans to a city at night with illuminated buildings and other structures. An illuminated U.S. flag appears near the bottom of the screen.
45–:52	and as for carbon dioxide, it isn't smog or smoke, it's what we breathe out and what plants breathe in.	There is a view of Earth from space with about ⅓ of Earth viewable, then the camera quickly pans to an aerial view of clouds moving. Then a woman blowing bubbles through a pink bubble maker appears, followed by a view of forest that pans up from the forest floor.
53–:55	Carbon dioxide, they call it pollution.	There is a close-up, ground-level view in a field with white and purple flowers. "Carbon dioxide" is printed in white on screen. Fades to black with carbon dioxide remaining on screen. The final shot is a young girl with pigtails blowing through a dandelion. At the end of the video, "Competitive Enterprise Institute" appears in white near the bottom of the screen. Beneath that is the URL www .cei.org, which is a link to the Competitive Enterprise Institute.
56–:59	We call it life.	

sectors, as well as transportation, technology, labor, and food and drug regulation. During the George W. Bush administration (2001–2008), CEI played a prominent role advancing climate change denial (Dryzek et al., 2011). In 2006, CEI launched a national television campaign by producing two commercials that promoted the benefits of carbon dioxide for the environment. This YouTube video, *Global Warming: Glaciers,* is one of these television commercials.

By February 2022, this video had garnered over 233,000 views, far fewer than the 10 million views with the Facebook video from Chapter 2, yet given its original life as a television advertisement, the number of those who have seen the content of this video is much greater than the number of YouTube views. In terms of the broader contexts of denial, the CEI functions much like the Heartland Institute as a key component of the "climate change denial machine" (Dunlap & McCright, 2011).

When John and Todd sat down to evaluate the reliability of the glaciers video, with no prompts or questions to directly address, the most salient aspect of what they brought to this deliberation was their academic or disciplinary perspectives. John noted how his disciplinary background in science and chemistry shaped his sense-making: "I think that's kind of why I wanted more data . . . if I can see maybe like, I don't know, that some concentrations of CO_2 in the atmosphere and how this will change or what those changes [might be] . . . that would be a little more believable for me." John held a view of reliability in which evidence-based scientific inquiry can yield verifiable facts, if not scientific truths. He also did not state or suggest that the scientific argument in the video was misleading or inaccurate.

Todd located himself in social studies with an emphasis in history, stating, "As a historian . . . I don't want to see if the source is true. I want to see who, [and] what is it trying to tell us." He also pointed to this academic identity in terms of the work that he perceives historians do. He stated, "If I was trying to convince somebody of my argument, this would be a very good kind of hook to draw them in, very simply, succinctly summarize the sort of dismissive belief." Todd defined reliability in terms of a source's usefulness or utility, as he described how he could *use* the source (based on that source's purpose) to convince others. He then made a connection between the work of historians and his priorities as a social studies teacher:

> As a social studies teacher, my job isn't to point my students to what is true. It's to point them to . . . this is reliable, this is not complete garbage . . . believe one side but some people believe, this is what other people believe, [these are] accurate representations of [each side]. Make your own decisions.

Todd maintained that reliability meant accurate representations of a perspective, in this case about climate change—that is, the glaciers video was a "reliable representation" of a dismissive perspective. In other words, Todd was contending that a source can be reliable if it accurately represented someone's view rather than it being factually accurate or whether the source (person or group) has expertise on the topic. To this point, Todd stated that he "does not consider if something is true" because he doesn't "work in the truth business." While this statement seems troubling, it does align with instructional approaches in history education where the goal is for students to simulate the work of historians. In these cases, concerns about source reliability and bias are less important than focusing on the evidentiary potential of a source (Barton, 2005). We return to this issue in the next section.

Again, Todd and John conducted this paired reading and analysis with minimal teacher intervention. They were not asked to respond to any detailed questions or prompts in their deliberation. In terms of criteria for what counts as an agonistic encounter, disagreement was normalized in this deliberation with John and Todd, yet there was less evidence that political emotions were expressed freely or that new identities or positions were cultivated (though discerning others' emotional engagement can be difficult). A different type of source, perhaps one about a polarizing politician's climate policies might more directly evoke emotional responses and connect with particular identity positions. Politeness norms and wanting the other person to feel comfortable can also operate in paired discussions like this.

We now wonder about ways to further build upon this deliberation to consider the potential for a deeper agonistic encounter as well as the value of other deliberation approaches. To begin, we consider to what extent their deliberation might have evolved differently if John and Todd were asked to work through the FLICC framework with the glaciers video.

"I'm Not in the Truth Business" and FLICC

The central claim of this 1-minute YouTube video is based on a logical fallacy, a deceptive assertion about the benefits of carbon dioxide to humanity. While it is true that carbon dioxide is essential to sustain life, this assertion ignores all the destructive and dangerous impacts from global warming when carbon dioxide levels rise to particular levels. In other words, carbon dioxide does affect plant growth but the negative consequences of heightened CO2 levels far outstrip the benefits (Cook, 2020a). So, the core claim of this video is, at best, an oversimplification, a well-worn science denial technique.

As with the climate denial Facebook video in Chapter 2, this YouTube video also employs the denial technique of cherry picking. The video cites two research studies from the journal *Science* as evidence to invalidate the climate science consensus: Greenland's glaciers are growing and Antarctica's ice sheet is thickening. These claims, however, represent a minor part of the complete body of accessible evidence. The vast majority of glaciers are shrinking, not growing, and focusing on the ways ice sheets can thicken in the middle ignores the rapid growth of ice loss at the edges (Cook, 2020a). Moreover, some of the scientists cited in these two research studies responded to this television ad, claiming it misrepresented their conclusions. Soon after the commercial aired, Curt Davis (2006), the director of the Center for Geospatial Intelligence at the University of Missouri–Columbia at that time, stated, "These television ads are a deliberate effort to confuse and mislead the public about the global warming debate. They are selectively using only parts of my previous research to support their claims. They are not telling the entire story to the public."

Using FLICC to guide an agonistic encounter might make it clear to someone with Todd's perspective that denial techniques are at work in this *Glaciers* video. This might especially be the case for Todd, who identified that the careful examination of claims and evidence was essential to social studies. We wonder to what extent Todd might agree or disagree with the findings from a FLICC analysis. These points of agreement or disagreement can also be investigated further with John and Todd and perhaps with a whole class, diving more deeply into texts and frameworks like FLICC. The same can be said for posing questions to John and Todd about stories-we-live-by tied to the *Glaciers* video. How might these two future teachers, for example, discuss stories like "humans are the center of existence whose well-being is more important than other species" or "nature is a resource for humans to exploit?" Would they agree these stories-we-live-by are present in this video? And, what might lead Todd, for example, to view these stories as a problem?

Other follow-up steps could also be taken. John and Todd could apply the FLICC framework to denial texts about other topics to again identify and understand how denial techniques operate across other socioscientific and politically divisive topics (e.g., the claim that vaccines cause autism). This could then lead to examinations of historical topics with more direct ties to social studies curricula in which different forms of denial are evident, including Holocaust denial, the conspiracy theory that asserts the Nazi genocide of Jews did not happen. Comparative studies of denial can be generative. Kahn-Harris (2018), in the book *Denial: The Unspeakable Truth,* in fact chooses climate change denial and Holocaust denial as the two primary topics to center his investigation of denial and denialism and to highlight how similar orientations, approaches, and techniques are at play across these topics. While there are significant cultural-historical, social, and political differences between the Holocaust and climate change, the broader point here is wondering to what extent the FLICC framework and engaging in comparative studies of denial might help students challenge the view that a social studies teacher's job "isn't to point my students to what is true."

We acknowledge Todd's argument about the value of cultivating a plurality of perspectives about topics or issues. We also understand the view that social studies teachers are not in the "truth business" relates to the perspective that source reliability can be less important or even unnecessary when conducting some historical inquiries. Yet, students need to learn that climate denial texts like the *Glaciers* video are rooted in falsehoods, not facts—that "the reliable representation of that perspective" is a clear example of climate denial. In this sense, social studies teachers are in the "truth business."

Agonistic Encounters and Seeking Common Ground

The above example highlights what a deliberation can look like with an agonistic starting point—two people working across distinct, if not opposing, climate belief orientations and across different academic or subject matter disciplines—where there seems to be a low possibility of final resolution or reconciliation (Mouffe, 2018). A starting point that differs from agonism is seeking common ground by exploring and identifying what might be shared values about topics, policies, or programs, whether, for example, it is about the right for all people to have access to clean water, the need to limit food waste, or the position that fossil fuel companies should not be involved in setting environmental policy.

In her work as a climate change communicator, climate scientist Katharine Hayhoe points out how she can find shared values with roughly 93% of the U.S. population, all but the people who represent the "dismissive" orientation based on Yale Climate Change Communication surveys. Hayhoe has found this ever-shrinking percentage of people too ideologically invested to seriously entertain the possibility of changing their views. Yet, those located on the dismissive end of the spectrum are not a monolithic group. Todd, in his willingness to support a transition from fossil fuels to renewable energy sources, for example, differs from others with a dismissive orientation. So, even those from the dismissive end of the spectrum might be able to identify shared values with people holding other climate views.

Deliberation and Reflexivity Across Other Differences

Whether the approach is agonistic or focused on finding and building upon shared values, educators can incorporate difference across other dimensions related to students' beliefs, backgrounds, identities, and experiences. With a goal to deepen students' understanding of climate change, teachers can guide students to consider how their environmental views, political stances, and other perspectives might enrich their meaning-making and enhance a deliberation.

Environmental attitudes

In addition to the Yale Climate Change communication survey, teachers can also bring in tools like the New Environmental Paradigm (NEP) Scale, a measure of environmental concern that has been used in many countries around the world (Dunlap & Van Liere, 1978; Dunlap, 2008). The scale is based on responses to 15 statements that either endorse the new paradigm (e.g., "Plants and animals have as much right as humans to exist" and "The balance of nature is very delicate and easily upset") or support the dominant social paradigm (e.g., "Humans have the right to modify the natural environment to suit their needs" and "Human ingenuity will insure that we do not make the Earth unlivable").

Other perhaps more informal approaches can also encourage students to be reflexive about how their environmental beliefs, attitudes, or behaviors inform their sense-making during deliberations. This can include their views about the natural world, animal life, the value of clean water and air and food quality, as well as concerns about food and water shortages and the impact of pollution on different ecosystems and communities, among a range of other topics, depending on the unit of study. Reflecting on these views helps students mobilize what they already care about, which can forge links to deliberations about climate change solutions, rather than fixating on whether climate change is happening or not.

Political perspectives

Because political identities and perspectives shape how we view and understand the world, attending to differences across different political perspectives can support deliberations. In the United States these differences are often framed reductively in dualistic terms: One is either a Republican or a Democrat, holding views that align with the respective political party, a reflection of what Ralph Nader has called an entrenched duopoly in the country. While a binary framing of Republican/Democrat fits with a typical debate structure, a more robust and comprehensive view of political perspectives to facilitate student deliberations can be used. Lee Drutman (2021) offers an online survey or quiz where readers answer 20 questions that target social and economic beliefs and values. The results place people in one of six new political parties: Patriot Party, Christian Conservative Party, American Labor Party, Growth and Opportunity Party, New Liberal Party, and Progressive Party. Echoing the arguments that public intellectuals and civic leaders like Ralph Nader have been making for decades, Drutman (2020) contends the current two-party system is toxic and "trapped in a doom loop of escalating two-party warfare from which there is only one escape: Increase the number of parties through electoral reform." Much like the Yale Climate Change communication survey, which outlines six different climate change profiles (rather than an either-or framing of believing or not believing in human-induced climate change), these six political parties (among others) expand the ways students can situate themselves in a classroom deliberation. There is also no need to require students to choose or align themselves with a political party or perspective. The larger goal is to enlarge the space of possible entry points into a deliberation.

Identities

In addition to mobilizing environmental and political perspectives, students can also draw on other beliefs, values, identities, and experiences to enrich classroom deliberations. Gender, geography or place, racial identity, religion, income level, dis/ability status, among others, offer additional lenses students might choose to mobilize to engage with climate-related texts. A young female Bengali student with extended family in Bangladesh, for example, can marshal particular insights and understandings about devastating climate change impacts as warmer temperatures cause the Himalayan glaciers to melt and sea level to rise. Of course, no student can represent "the" perspective for any group. Nor do we advocate that students hold on too tightly to any labels (Manji, 2019). The broader goal is to enlarge the classroom space for a plurality of perspectives, reflecting a commitment to a pedagogy of counternarration where students learn to focus on voices from the margins to examine and understand dominant and nondominant social and political discourses or narratives (Gibson, 2020). We expand on the idea of counternarration in the next chapter.

Emotions

Creating conditions for a range of emotional responses to be expressed in a deliberation is another way to engage with differences. Rather than stifling emotions, educators can encourage students to express how their views, understandings, or perspectives are informed by their feelings or passions (Barton & McCully, 2007; Pace, 2019). There is also clear evidence that youth are invested emotionally in climate change. One global research study led by the University of Bath in the United Kingdom, in collaboration with five other universities, for example, found that nearly 60% of young people (ages 16–25) felt very worried or extremely worried about climate change and two-thirds reported feeling afraid, anxious, and sad. Valuing students' feelings and emotions as a legitimate source of their sense-making, rather than creating a false separation between thought and feeling (Boler, 1999) can open up additional spaces for students to work with denial texts.

Moreover, emotions are part of students' sense-making when they are being reflexive across these different areas (climate beliefs, subject area backgrounds, environmental attitudes or values, political views, and identity markers). Motivated reasoning is an emotionally driven process where we tend to select and apprehend information to confirm what we already know or when we perceive a threat to our identity or moral framework. Moral judgments are made "quickly and emotionally" because "we do moral reasoning not to reconstruct the actual reasons why we ourselves came to a judgment: we reason to find the best possible reasons why somebody else ought to join us in our judgment" (Haidt, 2012).

AFFECTIVE LITERACY

In response to escalating ecological, social, political, and economic problems in the world, including climate change, Sharon Stein (2021) contends that global citizenship education needs to move beyond transmission-based content or competency approaches to emphasize critical and affective literacies to cultivate students' ongoing self-reflexivity, accountability, and discernment. Stein argues that "the challenges we face are not primarily intellectual problems that can be solved with more facts, information, or 'content'—they are also affective and relational problems" (p. 489). She encourages educators and students to engage in forms of reflexivity that examine not just our actions or our thinking, but to reflect on the relationships between behaviors, thinking, and emotions—to examine our habits, desires, and goals.

She calls for "affective literacy" to recognize and accept responsibility for the harmful impacts of our desires, investments, and emotions while learning to manage challenging emotions without being overwhelmed or immobilized. In the case of eco-civic deliberation and agonistic encounters this would mean staying engaged with each other despite our differences and the emotions they might elicit. This requires what Stein calls "relational rigor," the ability to be respectful of other viewpoints, include marginalized perspectives, and maintain trust while interrupting and addressing patterns of behavior, thinking, and emotion that have perpetuated injustice. While critique of existing systems and stories-we-live-by is necessary, we also need the reflexive capacities "to disarm our common defences and develop a suspicion about our own presumed innocence and the benevolence of our desires" (p. 490) in order to envision new possibilities for knowing, being, and relating.

Helping students understand the affective dimensions of their responses can enrich class deliberations and point to compelling inquiry questions to pursue, such as: How did I come to hold particular views about a topic like climate change? What emotions does this topic (or text) evoke? Why might I have these feelings? What family or community experiences or factors played a role? How does my information or media diet affect my thinking? Can I identify particular texts or sources that have had the greatest influence on me?

CONCLUSION

Deliberation and reflexivity are indispensable to posing and pursuing questions about the reliability of diverse, complex climate-related sources of information that students are likely to encounter. To promote civic dialogue and classroom deliberations, interpretations and understandings of texts need to be grounded in careful reflections of our own and others' views. This commitment to reflexivity facilitates awareness of the limits of our own views as we engage different perspectives and endeavor to determine which sources to trust and why. We can also continue to investigate questions about why people believe in particular denial texts as well as what might persuade them to challenge and confront climate denial. The next chapter considers this terrain with an exploration of reliability stories.

CHAPTER 6

Confronting Denial Through Counternarration and Reliability Stories-*To*-Live-By

In this chapter we look across our work with undergraduate students to identify how they appeal to different *reliability stories* to help determine who to trust and why. We begin with a model we developed to help secondary level students more skillfully evaluate websites. Next, we highlight how the work of undergraduate students like John and Todd, in Chapter 5, pointed to particular reliability stories the students were drawing upon as they evaluated climate change websites. Then, we mobilize the tools of counternarration to craft *reliability stories-To-live-by*.

DISCERNING RELIABILITY: A STARTING POINT

In previous work, we outlined key procedures and associated conceptual understandings for evaluating the reliability of sources (Damico & Baildon, 2015). We demonstrated that it is not enough that students follow literacy or critical thinking procedures; they need to understand *why* discerning reliability is important, *why* they are using these procedures, and *how* these procedures can be applied to determine reliability in real life. The left-hand column of Table 6.1 lists procedures for discerning reliability; the right-hand column describes core understandings required for each procedure. Three concepts are central to determining reliability:

- When considering source provenance, it is important for students to know that all authors are in some ways biased, yet that is no excuse to dismiss all sources as equally biased; authors' backgrounds, experiences, and willingness to be up front about their interests count as well.
- When considering purpose, it is important to understand that all sources are written with a purpose or agenda, but sources vary

in how transparent these purposes are, and some purposes and motivations are more trustworthy than others.

- In terms of content, the careful evaluation of claims and evidence and corroboration (cross-referencing) is key to gauging source reliability.

These three concepts interact dynamically, so an ultimate decision about the reliability of a source needs to be holistic, a careful assessment of the interdependence of these core concepts. Table 6.1's procedures and corresponding conceptual understandings outline how reliability determinations of any given information source need to be part of a holistic assessment of both text and larger conceptual understandings (see final procedural step).

Table 6.1. *Key Procedures and Conceptual Understandings for Evaluating Reliability*

Procedure	Key Conceptual Understandings
Evaluate provenance	1. Author background, expertise, experience affect their competence to speak about a particular issue (depends on issue they discuss and their experience with it). 2. All authors are biased or have limited views, but the bias must be evaluated to determine whether it is acceptable or if it should disqualify them. 3. There is a disclosure that states that one's background, interests, and positions are important. If not stated, can do background check.
Assess purpose	1. All sources are written/created with some purpose in mind. 2. Some purposes are explicitly stated while others may be vague, implied, hidden, or not easily discerned. 3. Some purposes are more trustworthy than others (e.g., inclusive, sincere, respectful, balanced purposes are more trustworthy than efforts to mislead, deceive, or solely promote self-interest).
Analyze content and cross-reference	1. Claims and evidence must be evaluated for accuracy. This requires evaluating reasonableness, whether content fits with what is already known, what other sources say, etc. 2. Content needs to be checked for errors, bias, and tone (e.g., if emotive, one-sided, advocacy-focused, etc.) to determine if reliable as a source of information. 3. Students need to corroborate information and check to see if it is consistent with other sources or if it is refuted by other sources.
Make determination of reliability	Each of the above factors must be weighed together to make an overall judgment of reliability, because informed conclusions and decisions require reliable information.

Source: Damico & Baildon, 2015.

As a recursive process, the procedures outlined in Table 6.1 offer clear steps to help determine the reliability of information sources as part of an inquiry. These procedures are similar to other models of inquiry in the social sciences (VanSledright & Afflerbach, 2005; Wineburg, 2001) and in disciplinary literacy (Spires et al., 2021). These kinds of approaches draw on critical literacies (Damico & Baildon, 2011) to examine specific claims and evidence and assess the relationships of source provenance, author purpose or motivation, the content of the source, and the landscape in which a text operates.

RELIABILITY STORIES-WE-LIVE-BY

In our previous work examining how undergraduate students (with varied climate beliefs and different academic backgrounds and areas of expertise) evaluate the reliability of climate change texts (which we discussed in Chapters 4 and 5), we have discerned three core "reliability stories" in terms of how they determined trustworthiness (Damico, Baildon, & Panos, 2018). We named these *reliability stories-we-live-by* due to the ways these stories circulate in the minds of individuals and in society. Like broader cultural stories-we-live-by, these reliability stories are often not easily identified as stories since they operate as taken-for-granted cultural assumptions and need to be analyzed critically and resisted if necessary (Stibbe, 2021).

The ways students deliberated the climate change sources pointed to three reliability stories: (1) having the "other side" represented leads to a more balanced understanding of climate change; (2) having more information or evidence is needed to support an argument; and (3) reasoning involves the evaluation of one's own identity and perspectives.

Reliability Story #1: Having the "Other Side" Represented Leads to a More Balanced Understanding of Climate Change

Across student deliberations, there was a perceived need that sources should include other views about the climate "debate" to be reliable. One pair of students, for example, stressed that opposing views had to be presented in a balanced way within a source and the "other side" needed to be presented fairly. They thought that evidence from "different sides" could be provided by "interest groups," others who have done studies and have data based on research (e.g., "professors"), along with people "on the other side of the continuum." For example, one student argued that it would have been better if the Competitive Enterprise Institute (CEI) source hadn't been dismissive of the opposing side's claims. Students thought opposing views had to be acknowledged, "even if you think it's wrong." They thought the climate change "debate" required a fair and balanced representation of "both sides" of the debate.

Wanting the "other side" presented in a fair, honest, or balanced way appears to be a laudable goal desired by these students as they deliberated the reliability of different video sources. The notions of fair and balanced information sources, making sure all sides are represented fairly, and the call for multiple perspectives are all stories that operate in broader society. This reliability story is baked into the very fabric of civic life in the United States in many ways: our two-party system, a focus on debate in classrooms, the polarizing binaries of "us versus them," liberal versus conservative, prolife versus prochoice, among others, and what Hess (2009) has called the "rhetoric of balance, fair play, and multiple perspectives" (p. 123).

Reliability Story #2: Having More Information or Evidence Is Needed to Support an Argument

A related story that emerged from the students' deliberations was that reliability depended on whether *enough information* was provided to support the main claims made by the video sources *rather than an evaluation of the credibility or accuracy of the information itself*. With the CEI glacier video, for example, one student emphasized the need to "get as much information as possible," and that "no matter what I read, I'm gonna dig deeper into it." One student desired information presented in a more "scientific manner," and stated that the source had to at least "sound" scientific, while another student wanted more quantitative data to support claims.

We find this to be a commendable reliability story, and one that also has parallels in our broader culture. People do often need more information, want to corroborate information they have, and have to determine when they have "enough" information to make an informed decision.

Reliability Story #3: Reasoning Involves the Evaluation of One's Own Identity and Perspectives

Another reliability story relates to ways students have referenced aspects of their identities to evaluate sources. For example, one student talked about being an art educator and evaluating the sources with an "artistic view" by drawing on particular aesthetic criteria, such as whether it was "stylish," "had better effects," and how it was "put together to make us feel a certain way." Students have also discussed the need for texts to "connect" to them as readers. This sense of connection and trust in sources has depended on what they already knew or believed. Of course, readers employ a range of identity resources (e.g., values, beliefs, background knowledge) in evaluating information related to gender positions, sexual orientation, cultural backgrounds, social class, race, nationality, religion, ideological perspectives, and professional identities (Damico, Baildon, Exter, & Guo, 2009). And, as the previous chapter highlighted, encouraging this type of reflexivity is notable.

Reliability Stories and Climate Denial

These three reliability stories possess a commonsense quality that appears useful for working with information sources: We want students to seek and appreciate different perspectives of an issue (including the "other side"), acquire more information or data to better understand an issue or to support their arguments, and reflect on their own social locations as readers. Yet these reliability stories are also problematic for the ways they can, even inadvertently, promote and spread climate denial.

Again, making sure "the other side" is included is not useful if the other side is based on erroneous information or when the goal is to achieve an objectionable "balance." Climate science denial texts produced and spread by the climate denial machine (Dunlap & McCright, 2011) want citizens to believe there is a legitimate "other side" to the "debate," despite the scientific consensus. This type of false balance or false equivalency can be common in journalism to promote "fairness" by giving the "other side" an equal platform to present baseless claims. In their study of journalistic reporting of climate change, Brüggemann and Engesser (2017) found that while the scientific consensus on global warming has grown more solid in recent decades, public opinion has remained uncertain about climate change as a problem, especially in the United States, the United Kingdom, and Australia. They attribute this to the media providing a forum for contrarian views that promote uncertainty about climate change.

Similarly, the acquisition of more information to support an argument is not worthwhile if the argument is faulty or the information untrue. Again, denial texts like CEI's glaciers video suggest we need more information, that "the jury is still out" or that there is a body of evidence that climate scientists have not fully taken into account, such as information contained in the reports about glaciers and the Antarctic Ice Sheet. Moreover, it is easy to locate additional information to support just about any perspective or argument, much of it readily available online. For example, many people get their information about a wide range of topics, including climate change, from social media such as Facebook. The social media giant has become a primary target for environmental and science-based groups, including Friends of the Earth, 350, Greenpeace, Sierra Club, and the Union of Concerned Scientists, who sent a letter calling on CEO Mark Zuckerberg to stop Facebook from spreading misinformation that threatens the ability of citizens and policymakers to fight the climate crisis. While information has never before been so abundant, simply seeking more information is not necessarily what is needed to better understand or address climate change. As Oreskes (2019) argues, climate deniers create the impression that the science is not settled and if "we respond on their terms—offering more facts, insisting these facts *are* facts—then they win, because now there *is* contestation" (p. 246, italics in original).

Finally, it is not enough for readers to acknowledge their backgrounds and perspectives given how easy it is to read *with* one's own perspective and social locations. We are all motivated readers, which means we come to texts with particular identities, ideas, values, experiences, and positions that encourage us to receive and accept information in ways aligned with our pre-existing beliefs and perspectives (Damico et al., 2020). We need to be especially careful of determining information that supports what we want to believe as most reliable. Thus, these stories, like all stories we tell ourselves, merit scrutiny.

COUNTERNARRATION

Concerns about the three reliability stories can be addressed with the tools of counternarration. Melissa Gibson (2020) highlights how processes of deliberation can privilege certain types of discourse and power relations in classrooms that work to ignore, dismiss, or exclude marginalized voices. She calls for counternarration to expose taken-for-granted stories that perpetuate injustice. In the context of climate denial, taken-for-granted reliability stories can delay necessary action.

Industry-led and politically motivated denial has tapped into deep cultural narratives that have perpetuated injustice and marginalized whole communities most vulnerable to climate change, while using their vast financial resources and political influence to spread misinformation, disinformation, and delay timely action. Counternarration helps question "where we turn for knowledge, who we allow to produce knowledge, and what we consider worth knowing" (Gibson, 2020, p. 442). Counternarration can help mark a transition from *reliability stories-we-live-by* to enact *reliability stories-To-live-by*.

Reliability Stories-*To*-Live By

To outline new reliability stories, it is necessary to consider the nature of denial texts and broader contexts of denial, including the misinformation platforms granted to corporations and powerful elites (Monbiot, 2017; Nichols & McChesney, 2010; Warzel, 2018) that have distorted public information, degraded public discourse, and eroded social trust and democracy (Baildon & Damico, 2019; Kavanagh & Rich, 2018). Table 6.2 highlights how counternarration can be used to revise the three limiting and potentially troublesome reliability stories-we-live-by outlined above and to craft three *reliability stories-To-live-by*. To do this we consider notions of public or societal trust and conclude with an initial thought experiment about a "Trust for Public Information." Building on the work of the Trust for Public Land, there could be two key components of a Trust for Public Information: First, like the environment and public lands, information needs to be protected and skillfully managed on behalf of citizens for common or public interests rather than for private gain. This includes protection from misinformation and disinformation that promotes ecological, societal, and civic harm. Reliable, trustworthy information needs to be safeguarded and cared for. Second, a Trust for Public Information would be shared and held in common by present and future citizens, understood as a vital resource for communities to access, use, and contribute to in order to advance public health and civic well-being.

Climate Science Is a Well-Established Knowledge Base That Merits Public or Societal Trust

This is a straightforward *reliability story-To-live-by*. It requires acknowledging and accepting the results of Western scientific research that dates to the mid-19th century and the cumulative knowledge and wisdom from non-Western and Indigenous modes of scientific inquiry. Chapter 1 summarized the climate science consensus and considered reasons why people should trust this consensus. The primary reason is because this knowledge base is grounded in rigorous methods to collect and analyze data in which findings are critically vetted in communities to ensure claims are substantiated with evidence and errors are corrected (Fish, 2001; Oreskes, 2019; Sinatra & Hofer, 2021).

The tool of counternarration comes into play with this *reliability story-To-live-by* to help counter climate denial. While Western-based science and the voices of climate scientists working in this tradition are not typically marginalized (in ways that call for counternarration), the climate denial machine has succeeded in getting more people to question the legitimacy of climate science in part by targeting climate scientists and impugning their integrity and credibility. Thus the need for this *reliability story-To-live-by* that directly counters industry-led and politically motivated climate denial and the cultural-historical narratives that have had destructive ecological and human consequences. Finally, this *reliability story-To-live-by* impugns the need to give equal weight or value to "the other side." While some uncertainty, along with human fallibility is baked into the scientific process, which requires ongoing scrutiny and correction as the knowledge base grows, there is no legitimate "other side" in terms of climate science. There is either a consensus or there isn't. There is.

The Type and Quality of Evidence Determines the Soundness of an Argument

It is not just the need for "more information" that promotes reliability; it is a need for more trustworthy information to confirm or contradict one's understanding. In many ways, this story serves as an antidote to science denial made visible with the FLICC framework. Rather than "fake experts" we depend on well-established experts with requisite knowledge and mastery of the relevant material; instead of "logical fallacies" we look for carefully reasoned explanations; rather than succumb to "impossible expectations" we operate with realistic projections and sensible agreements in mind; rather than "cherry picking" data, we appeal to the preponderance of evidence standard to best represent what is known about a topic; and instead of embracing "conspiracy theories" we again work with the best available evidence that is based on systematic and rigorous inquiry.

Rather than represent the particular "private" interests of industries and corporations shaped by the financial need to maximize profits, this *reliability story-To-live-by* emphasizes the need to seek information that is in the interests and needs of the wider public. While the two us would not claim that some information can be neutral or value-free, there remains a stark difference between information presented by climate denial organizations like the Nongovernmental International Panel on Climate Change (NIPCC) and information from the Intergovernmental Panel on Climate Change, the United Nations body to assess climate change science. In this sense, the IPCC offers information, reports, and resources that are in the public trust: This information has been expertly managed on behalf of citizens, rather than packaged to promote private interests. The information is also a resource communities can use to guide democratic dialogue and decision-making for climate action.

As with the above *reliability story-To-live-by*, this one also requires active counternarration to challenge all the ways the climate denial machine and related political ideologies and projects have attempted to erode faith in public institutions and weaken confidence in experts, expertise, and established knowledge (Nichols, 2017). An emphasis on how the *type and quality of evidence determines the soundness of an argument* also calls for a broad conception of what counts as evidence and "established knowledge" to include the voices, experiences, and cumulative wisdom of those who have been invisible, silenced, or on the margins of public discourse and debate about causes, consequences, and potential solutions for climate change—namely, women, communities of color, and those disproportionately affected by climate impacts.

Reasoning Involves the Critical Evaluation of Varied Perspectives, Responsibility, Complicity, and Privilege

A third *reliability story-To-live-by* is one in which people are aware of their identities, emotions, perspectives, and social locations yet go beyond them to engage with others' views and lived experiences. Counternarration, in this case, can help individuals of relative privilege become more aware of this privilege to discern how particular narratives, beliefs, values, and lifestyles they have adopted (carbon intensive marked by high consumerism) have been contributing to ecological destruction and injustice. Counternarration serves the vital role of listening to the voices of those most affected by the negative impacts of climate change, while exposing the complicity of those who benefit from those structures (Global North, U.S. citizens, the wealthiest one percent of people around the world, etc.). Instead of reasoning through one's

Table 6.2. Counternarration for *Reliability stories-To-live-by*

Reliability stories-we-live-by	Limitations	Key questions	*Reliability stories-To-live-by*
Having the "other side" represented leads to a more balanced understanding of climate change	Two-sides framing can be a logical fallacy and create false equivalency that valorizes extreme minority views	Is the "other side" based on reliable, accurate information and evidence? What are the purposes or agendas of each side?	Climate science is a well-established knowledge base that merits public or societal trust
Having more information or evidence is needed to support an argument	Self-sustaining reference system of online information; facilitates denial techniques like cherry picking	What information do we need? Whose information should we trust? What standards of evidence will we use?	The type and quality of evidence determines the soundness of an argument
Reasoning involves the evaluation of one's own identity and perspectives	Not enough for readers to acknowledge their backgrounds and perspectives given how easy it is to read with one's own perspective and social locations	What influences my understanding? What are the limits of my perspectives? What might complicate my own understanding? Whose voices are left out?	Reasoning involves the critical evaluation of varied perspectives, responsibility, complicity, and privilege

own identity and perspectives, a reflexive stance requires observing one's cognitive and emotional responses to texts in order to recognize the ways one's particular perspectives are potentially limited, harmful, and reproductive of existing systems of injustice (Stein, 2021). This requires intellectual humility and a "relational rigor" to decenter one's own views, ensure inclusion of marginalized voices, and develop relationships and social trust with others as a basis for rethinking reliability (Stein, 2021).

A TRUST FOR PUBLIC INFORMATION

The Trust for Public Land (TPL) is a nonprofit organization founded in 1972 to help ensure all people in communities and cities have access to the outdoors and nature as a matter of health, equity, and justice. TPL "creates and protects land for people, ensuring healthy, livable communities for generations to come" (www.tpl.org). For Brendan Shane, climate director for TPL, cultivating relationships is core to the work of TPL as they partner with a range of community stakeholders (e.g., youth, schools, builders, landscape architects) to help connect people to nature and to each other. This aligns with a core civic mission to create parks and green space in ways that build community trust, social cohesion, and improve a sense of belonging. Shane (2022) also notes that climate change requires keeping "long-term stewardship in mind" as TPL increasingly comes to view its mission to protect land as an "ongoing commitment to help communities respond to climate impacts," such as sea level rise on a recently protected coastal ecosystem.

TPL is a springboard to consider similarities between the protection of land and green space with the protection of reliable information: to think about how, for example, a "Trust for Public Information" might help promote *reliability stories-To-live-by*. Since a more extensive exploration of this idea is beyond the scope of this book, what we offer here is more a thought experiment than a concrete proposal.

A Trust for Public Information starts with the premise that reliable information needs to be protected and well managed in the interests of the public good. Central to these efforts is protection from social and ecological harm wrought by misinformation and disinformation. Trustworthy information needs to be conserved and defended. Much like the Trust for Public Land, a Trust for Public Information could facilitate policies and practices within and across communities to help ensure that reliable, trustworthy information is accessible to guide sound decision-making about any number of issues in the public interest. Of course, there is a significant difference between protecting land and safeguarding information; land is finite, information continues to expand at exponential rates. This is why it is crucial to not rely on a single organization to do the work of information protection and management. Rather, protection from mis- or disinformation and promotion of trustworthy information is a shared responsibility among all citizens, a responsibility that is attuned to local contexts as groups and communities hold each other accountable to help improve public and civic health within and across settings. The critical literacies and eco-civic practices outlined in this book can help cultivate accountability and responsibility for individuals and groups. And, of course, classrooms would be communities to learn about and practice ways to access and advance the public trust. Ultimately, just as there is a need to protect land and people's relationships to land, there is a need to protect reliable information and people's relationships to reliable information.

In fact, a variant of this idea has already been advanced. In April 2022, the Disinformation Governance Board, a new advisory board within the United States Department of Homeland Security (DHS), was created to deal primarily with national security threats tied to misinformation and disinformation spread by "foreign states such as Russia, China, and Iran" as well as "transnational criminal organizations and human smuggling organizations" (Fact Sheet, 2022). Given the significant backlash to the formation of this board for its perceived threat to free speech, its future is uncertain. While the idea of a Trust for Public Information differs from this Disinformation Governance Board (for starters, it is important to not rely on a single, centralized institution), it is clear that navigating free speech concerns will likely be central to advancing any initiatives related to a Trust for Public Information.

CONCLUSION

Reliability stories-To-live-by are grounded in understandings of the web of denial contexts, the climate denial machine (Dunlap & McCright, 2011), the social organization of denial (Norgaard, 2011), and the ways these broader conditions shape our sense of trust, identity, and knowledge. *Reliability stories-To-live-by* are premised on the notion that reliable information is necessary in a democratic society in order to make decisions based on reality rather than on disinformation and falsehoods. The idea of a Trust for Public Information invites additional exploration.

CHAPTER 7

A Deeper Dive Into Climate Denial
Classroom Inquiries

"In nature's economy the currency is not money, it is life."

—Vandana Shiva, *Earth Democracy: Justice, Sustainability, and Peace*

Climate denial is arguably the most consequential topic of our time. A failure to directly confront denial and respond comprehensively to the climate crisis with urgency and care will likely lead to catastrophic species and habitat loss across the planet. Climate crisis and denial challenge political, social, and economic systems to manage extreme weather events, disruptions to basic services, and security issues. These effects will reverberate across the world and continue to be felt the most by vulnerable populations, those who have contributed the least to the crisis. This is why climate denial needs to become a curricular and instructional priority in schools and other sites of learning across all age levels. As the preceding chapters have highlighted, the content areas of literacy and social studies (with their corresponding sets of tools, frameworks, and ideas) are ideally suited to this effort. Critical literacies+ along with eco-civic practices of deliberation, reflexivity, and counternarration can help students identify techniques in climate denial texts, situate these texts in larger contexts of denial, and discern ecologically destructive stories-we-live-by. And Chapter 6 charted a needed transition toward *reliability stories-To-live-by*. Confronting climate denial depends on educators and students using the tools of reflexivity and counternarration to help lead this transition. Ultimately, when it comes to climate change texts or sources, questions about their reliability and trust are bound inextricably in content understandings of denial contexts.

This chapter builds from Chapter 6 to consider additional ways literacy and social studies educators can make climate denial more central to their work. This includes approaches to more fully reckon with destructive stories-we-live-by and to explore *ecojustice stories-To-live-by*. Ultimately, as educators committed to confronting climate denial, we not only need to understand how denial works; we also need to offer learning pathways that lead to inquiry-based explorations of more just and sustainable futures, what moving past or through climate denial can look like.

INQUIRY PATHWAYS: DENIAL CONTEXTS AND
STORIES-WE-LIVE-BY

As we have demonstrated, climate denial texts are shaped by particular political, social, economic, and cultural stories-we-live-by including: Humans are the center of existence and separate from nature; the goal of a society is perpetual economic growth without limits; and nature is solely a resource to be used for human purposes (Stibbe, 2021). These interrelated, mutually reinforcing stories influence how we think about our social, economic, and political systems and institutions; concepts like progress and freedom; the environment; and ourselves. These stories are also destructive: They work against the well-being of all species, reinforce unsustainable patterns of production and consumption, and set up relationships between people and nature that are divisive and alienating (Stibbe, 2021).

The fossil fuel industry and the climate denial machine have promoted these stories-we-live-by while gaming, neutralizing, and undermining democracy to ensure that political power remains in fewer and fewer hands (Freese, 2020). Comprehensive, highly coordinated, and well-financed misinformation and disinformation campaigns have distorted and debilitated trust in reliable and trustworthy information sources and have led to a significant delay in effective climate action. As outlined in Chapter 3, these stories are also steeped in a web of overlapping denial contexts that align with academic disciplines and subject matter areas in social studies—economics, political science (civics), history, psychology and sociology, media, and geography. So this is a reasonable place to begin outlining some potential questions that educators can pursue with students. Yet, as will become clear, many of the inquiry questions for each subject area overlap and intersect, thus inviting a more inter- or multidisciplinary perspective. These questions could also be revised or developed further through the inquiry design model (Swan et al., 2018). Of course, this is by no means an exhaustive list of potential inquiry questions; many more possibilities exist.

Economics

Who is most responsible for climate change denial? This question invites students to explore the climate denial machine led by fossil fuel corporations in concert with organizations such as the American Petroleum Institute and conservative foundations like the Competitive Enterprise Institute and the Heartland Institute. Students can investigate in what ways these organizations enlisted scientists, economists, and politicians in their efforts to promote fossil fuel interests, influenced energy and economic policymaking, and supported media campaigns to spread climate denial messaging. Analysis of the range of techniques and tactics used in denial efforts also offer potential inquiry pathways. Students can compare industry-led climate denial with other efforts to deny well-established evidence about social harms, such as tobacco industry campaigns. This question also provides opportunities for students to consider the role of traditional media and social media companies in spreading denial and climate misinformation, the role of governmental in-action and deregulatory economic policies, and the influence of dark money donor networks to promote particular economic agendas, including the role of denial in these agendas (Mayer, 2016).

Can an economic system of capitalism adequately address climate change? This inquiry question provides an opportunity for students to ex-plore in what ways capitalism as an economic system and the role of private property, profit-maximization, markets, competition, consumerism, and the focus on growth and GDP have contributed to climate change and denial. Drawing on the C3 Framework for the Social Studies State Standards (NCSS, 2013), students can examine the ways individuals, businesses, governments, and societies make decisions based on these factors and the costs and benefits (i.e., trade-offs, externalities) of those decisions in terms of economic growth and climate change. This inquiry question also enables students to investigate the ways the capitalist system (e.g., markets, investment, human and physi-cal capital, and consumption) is already transitioning in key economic sec-tors to address climate change. For example, transportation is moving away from gasoline-powered engines to electric cars, buses, and trucks; there are efforts to shift energy markets to solar, wind, and other zero-emissions energy sources; carbon markets provide carbon pricing and trading mechanisms, such as cap and trade, to offset emissions; industries and organizations are reducing carbon intensity by using more energy-efficient building materials and improving resource management; and many companies are engaging in carbon removal, capture, and storage efforts (Doerr, 2021). Students can ex-plore the ways economic policy and politics, investment and innovation, and citizens and consumers are contributing to these transitions in different sec-tors of the economy as well as consider the limitations of these efforts.

A related question is, To what extent are regulatory measures necessary to address climate change? Students can examine the efficacy of specific regu-lations, such as the Clean Air Act or the Clean Water Act, and Environmental Protection Agency regulations, as well as efforts by some politicians and in-dustry leaders to roll back or dismantle environmental regulations governing clean air and water, public lands, and toxic materials. This type of question can also be situated in a broader inquiry into neoliberalism: Is neoliberalism (as an economic and political project) irreconcilable with successfully ad-dressing climate change? Neoliberalism has driven national economic and political systems for the past few decades. It is a story of limited government, deregulation, privatization, the expansion of consumerism and markets into all aspects of life and has spread notions of progress as the accumulation of wealth. Students can investigate elements of this story, such as how pro-gress is conceptualized, impacts on policy, or how it has accelerated the impacts of human activity on ecosystems. Studying neoliberalism and the Great Acceleration (dramatic increase in human activity trends since the mid–20th century) provides an opportunity for students to examine relationships between socioeconomic trends (GDP, energy use, population growth, transportation, etc.) and Earth System trends (carbon dioxide, methane, deforestation, surface temperatures, etc.). The question, *Does neoliberalism have a future?* might be rich terrain for inquiry since continued impacts at these rates are untenable. This type of inquiry could also lead students to explore the possibility of nationalizing the U.S. fossil fuel industry in which the federal government purchases controlling ownership of U.S. oil and gas corporations like ExxonMobil, Chevron, and ConocoPhillips (Pollin, 2022).

Students also need opportunities to explore alternative economic models for confronting climate denial and combating climate change. Potential inquiry pathways include historical studies of Keynesian economics, a capitalist model with substantive government intervention during and after the 1930s Great Depression. For example, students might investigate New Deal environmental programs like the Civilian Conservation Corps and the Soil Conservation Service, which created jobs through conservation projects. There are also contemporary socially democratic national economies to investigate. Students could examine the economic and environmental policies of several developed and democratic countries that have much smaller per capita emissions than the United States, such as Finland, Denmark, Norway, Iceland, and the Netherlands (Aronoff, 2021). These countries are also notable for having high rankings in the 2021 UN World Happiness Report.

Other economic models for students to investigate are models of green growth, degrowth, and postgrowth. One generative question students might explore is, *Can we have prosperity without growth?* (Cassidy, 2020). This question would enable explorations of economic conceptions of "small is beautiful" and "economics as if people mattered" (Schumacher, 1973); green growth (continued economic growth based on sustainable development); degrowth, based on social and ecological justice instead of infinite growth and consumerism as primary economic goals (D'Alisa et al., 2015); and postgrowth, emphasizing collective and planetary well-being within ecological limits (Jackson, 2021; Post Growth Institute, www.postgrowth.org/). These models, along with others, such as "barefoot economics" (Max-Neef, 1992) and "doughnut economics" (Raworth, 2017), would also help students consider different ways to understand basic economic concepts like growth, prosperity, and development.

Students can also investigate the sustainable economic practices of Indigenous people in the United States or in other parts of the world. Students might investigate the question, *How do Indigenous economic models help address climate change?* to consider the ways particular cultural values and intergenerational, place-based ethical systems, forms of knowledge and social practice (e.g., traditional resource management, biodiversity conservation), and sustainable livelihoods might provide economic models for more sustainable futures.

Political Science and Civics

What has been the role of government in promoting climate denial? Students can investigate the role of the executive, legislative, and judicial branches in promoting climate denial. They can also examine the role of the Republican Party and the Democratic Party in promoting climate denial as well as the ways lobbying, political action committees, special interest groups, industry associations, and campaign financing have influenced policymakers. For example, students can investigate the ways these efforts have resulted in corporate-friendly energy, economic, and regulatory policies that constitute denial responses to climate change.

How do stories of freedom support or combat climate denial? With this inquiry question students explore the concept of freedom and the different views and stories related to freedom that characterize U.S. politics. For example, freedom from government regulation, freedom from limits, freedom of expression (to spread misinformation), freedom to accrue massive wealth and consume, and freedom from responsibility are all central themes related to climate denial. There are strong ideological commitments to freedom grounded in neoliberal, libertarian, and laissez-faire views that can be explored. There are also conceptions of freedom based on ideas that a comprehensive system-wide shift away from fossil fuel-based corporate capitalism toward a more social and democratic green economy could facilitate more autonomy for workers and citizens in the form of job guarantees, shorter work weeks, more time for low-carbon activities. In the book *Overheated*, Kate Aronoff (2021) cites the Green New Deal as an example to highlight how government intervention could "ensure our freedom to breathe clear and unpolluted air, to find a new home when ours floods or catches fire, to experience joy and contentment, and to live on a habitable planet." Aronoff contrasts this freedom with the freedom promised by neoliberalism, which is directed toward the one percent that have "the freedom to break up unions and pollute unimpeded, from regulations, and, perhaps most importantly, from democratic oversight."

Why are climate change and denial politically divisive issues? This inquiry question invites students to investigate the rise of political sectarianism in the United States, especially about climate science and climate denial. Related to the inquiry question posed in Chapter 3, *What best explains the range of climate change beliefs in the United States despite the overwhelming scientific evidence and consensus about the existence and causes of human-caused climate change?* Students might explore not only why there is a range of climate change views, but why climate change and denial have become partisan issues, divisive along political or ideological affiliations, how particular views are related to people's sense of political identity, and how social media has inflamed these divisions.

Another line of inquiry might explore the impact of government policy with a question like, *Is the U.S. Government more committed to promoting private (corporate) or the public good?* For example, students can examine whether public lands or clean water should be protected as public goods that benefit all of society. This question can also be explored historically and/or with more-recent case studies. For example, the tremendous wealth gains of U.S. billionaires during the COVID-19 pandemic tilt toward private wealth as a higher priority (Collins, 2021; Peterson-Withorn, 2021).

Another potential line of inquiry related to policy proposals considers who might be included or excluded from devising climate change solutions. The World Health Organization's (WHO) rules on global tobacco legislation might be instructive here. Article 5.3 of the WHO Framework Convention on Tobacco Control reads: "In setting and implementing their public health policies with respect to tobacco control, Parties shall act to protect these policies from commercial and other vested interests of the tobacco industry" (World Health Organization, 2003, p. 7). This type of document, which excludes tobacco companies from policymaking, operates on an international level, and applies to countries that sign the treaty, could be imagined as a "firewall for climate policymaking, applying not just to the big polluters but to the lobbyists paid to represent them" (Michaels & Ainger, 2020, p. 173). It is sobering to realize that a full-scale rapid transition away from fossil fuels needed to limit global warming would mean that trillions of dollars' worth of fossil fuel reserves could not be burned. This would "represent the single largest evaporation of private wealth since the Emancipation Proclamation" (Aronoff, 2021). Given that the fossil fuel industry wants to maintain its seat at the table and be viewed as integral to climate efforts and decision-making, it is worth exploring on what grounds they should or should not be involved.

Other related questions could explore the role of the U.S. government, with participation of other governments around the world as well as international organizations and associations, in enacting climate change policies and legislation. Inquiry pathways could include an examination of the role the United Nations has played in transnational efforts to deal with climate change, such as their establishment of the UN Framework Convention on Climate Change that has served to hold the annual Conference of the Parties (COP) to support climate action. Students can investigate the extent to which these efforts have been successful in supporting effective climate action. Specific multilateral climate agreements, such as the Kyoto Protocol (1997) or the Paris Agreement (2015), offer potential for analytic or evaluative inquiry questions (Swan et al., 2018) for students to examine the outcomes of these agreements.

National and international nongovernmental organizations or nonprofits committed to climate action, like Project Drawdown, Greenpeace, 360.org,

C40 Cities, Citizens Climate Lobby, Climate Action Network, Climate Justice Alliance, Friends of the Earth, youth-led organizations like the Sunrise Movement, and many other organizations engaging in climate action at local, state, national, or transnational levels, all offer opportunities for students to investigate the ways citizens all around the world are mobilizing and taking action to address the climate crisis.

Specific climate actions also offer inquiry pathways for students. Some examples include protests to stop the Line 3 Pipeline from the tar sands in Alberta, Canada to Wisconsin (https://www.stopline3.org/); Standing Rock protests against the Dakota Access Pipeline to protect Indigenous rights, access to water, and important cultural sites (Hayes, 2016); actions in Halkidiki, Greece, against the proposed expansion of gold mining in their region by the Canadian mining company Eldorado Gold (https://vimeo.com/146445852); opposition to the East African crude oil pipeline from Uganda to Tanzania (Nakate, 2022); or the Scientist Rebellion's demonstrations by over 1,000 scientists from 25 different countries calling for climate justice and immediate political action to address climate crisis (Osborne, 2022). Within and across a range of examples, students can inquire into what forms of political or civic action are effective in bringing about change.

History

What are the historical contexts of climate change and denial? This can include exploring the paradigm that was pioneered by the Royal Society and Francis Bacon in the 17th century about science and progress as a machine, humans as masters of the earth, and the earth's resources as serving economic growth and development (Klein, 2014; Lewis et al., 2015). This narrative accompanied the rise of capitalism, colonialism, and the need to extract carbon-based fuels to support revolutions in transportation, industrialization, and urbanization. Students can also study the history of the enclosures of common lands during feudalism. Accompanying these narratives of growth and "progress" are denials of the theft of land from Indigenous people and destruction of the environment.

Was the climate crisis avoidable? This is an evaluative question (Swan et al., 2018) where students can explore the potential for different historical outcomes. They could investigate a range of historical documents to learn more about what happened from bipartisan agreement in the 1990s about the need to address global warming to the growing doubt about the science and eventual staunch divisiveness. This question would support inquiry into different pathways of denial that could be traced as responses to the Club of Rome and its 1972 report, *The Limits to Growth* (Meadows & Randers, 2013), and major international climate conferences and agreements since (1992 Rio Earth Summit; 1995 Kyoto Protocol; 2007 Bali Action Plan; 2015 Paris Agreement, etc.). Students can answer this compelling question in different ways. One might, for example, argue that the current crisis was unavoidable given the production and consumption gears set in motion during the Industrial Revolution. Or one might contend the crisis was avoidable if federal and state governments acted in a more timely way to the available climate science and consensus about climate change known several decades ago. With this question, teachers might need to guide students to consider different historical turning points, from major shifts, such as the Industrial Revolution, to a range of particular events that offered early warnings about ecological loss and devastation (e.g., the publication of *Silent Spring* by Rachel Carson, the U.S. President's Science Advisory Committee report in 1965 that outlined concerns about a "greenhouse effect," the work of the IPCC beginning in the late 1980s).

What historical narratives (or cultural assumptions) are central to understanding climate denial? This inquiry question allows students to explore some of the main stories-we-live-by outlined in this book, along with core concepts such as progress, freedom, liberty, security, the good life, rights and responsibilities. Students can explore the consequences of these stories and assumptions, whether they might be considered destructive or beneficial, for whom, and in what ways. Inquiries into abolitionist movements of the past might also prove to be generative. Stephenson (2015), for example, makes a case that contemporary climate justice efforts align less with previous environmental campaigns and much more with the slavery abolitionist movement of the 19th century in terms of clear moral, political, and economic parallels between the two.

In what ways have religious or faith traditions and communities contributed to and/or resisted global warming? Stories like humans are the center of existence and nature is solely a resource to be used or exploited are supported and reinforced in some religious communities, which interpret sacred texts, such as the Bible, as confirmation of (hu)man's dominion over all other species on earth. Yet major religious leaders, including Pope Francis of the Catholic Church, have argued forcefully about the need to end our dependence on fossil fuels in the name of social and ecological justice (Pope Francis, 2015). Faith leaders across the world's religions, including Christian denominations, Sunni and Shi'a Islam, Judaism, Hinduism, Sikhism, Buddhism, Confucianism, Taoism, Zoroastrianism, and Jainism, have also joined Pope Francis in support of the ambitious goal to limit warming to 1.5 degrees. There are many possible entry points for exploring the ways religious institutions and specific religious communities are responding to climate change.

There is no shortage of other inquiry possibilities based on historical perspectives. In addition to investigating industry-led denial across the decades in books like *Industrial-Strength Denial* (Freese, 2020) and *The Merchants of Doubt* (Oreskes & Conway, 2010), students might also examine relationships among industry, public relations (PR), and the environment, in particular the ways PR in the United States since the beginning of the 20th century has shaped environmental issues and crises and how to respond to them (Aronczyk & Espinoza, 2021).

Psychology and Sociology

How do mechanisms of climate change denial work? Students can examine the role of motivated reasoning, confirmation and disconfirmation bias, identity-protective cognition, and the backfire effect (initial beliefs are reinforced when confronted) in climate denial. It is also an opportunity to explore denial as a psychological process, how it operates in individuals as well as socially. Students might explore media messages of "climate doom" and climate optimism and reflect on how these make them feel about climate change and the future: whether these messages make them more concerned (or hopeless) about climate change, and to what extent these messages motivate them to take action.

    How is climate change and climate denial affecting society? Students can investigate the social impacts of climate change and climate denial in terms of how they intersect with inequality or racism, for example, and their disproportionate effects on vulnerable populations. The societal impacts of climate change and denial include social and affective polarization (Finkel et al., 2020), social distrust, food insecurity, climate migration, and threats to cultures and livelihoods. A variant of this question is: *How is climate denial and inaction affecting well-being?*, which can spur investigations into climate or eco-anxiety. This can lead to questions like *What language might be most useful to help people navigate the psychological and emotional landscape of climate change?* For example, *eco-anxiety* is a term that helps define one's ongoing fear or worry about the state of the natural world or about the harmful environmental impact of one's actions, while *ecological grief* helps name the sorrow or mourning tied to current and future climate-related losses. Related to these terms is *solastalgia*, a word to capture the sense of loss and desolation tied to a place, home, or territory that has been ecologically altered (Albrecht, 2019). Along with solastalgia, Albrecht (2019) offers a lexicon "for a new world" to map a transition from the age of the Anthropocene, an epoch marked by humanity's significant impact of the Earth's ecosystems, to what he calls the "Symbiocene," a new epoch in which all forms of human action are intended to enrich mutual interdependence and mutual benefit for all species and the health of all ecosystems. Albrecht also offers the term *sumbiocracy* to name a new political system where all forms of governance remain committed to mutually beneficial relationships within a larger sociobiological system. Books like *Earth Emotions* (Albrecht, 2019) and *An Ecotopian Lexicon* (Schneider-Mayerson & Bellamy, 2019) can be source material for students to explore the affordances and constraints of language.

    Additional lines of inquiry include investigating claims and proposals about differences between the relative value and impact of individual versus collective actions in responding to the climate crisis—differences, for example, between lowering one's carbon footprint by reducing air travel, eating locally sourced foods, driving an electric car, turning off lights when leaving rooms, and more collective efforts such as organizing to oust fossil fuel friendly politicians and getting large-scale institutions to divest from fossil fuels. (The growing fossil fuel divestment movement is also a robust inquiry topic.) While confronting climate change is ultimately not an either-or issue, an individual–collective lens can be generative. This lens can also be used to analyze attempts by corporations and lobbying groups over the years to individualize social problems and deflect corporate accountability, with the tobacco industry and the gun lobby as two prime examples (Mann, 2021).

Media

What role has the media played in spreading climate denial? This inquiry question offers opportunities for students to examine how our media ecosystem (both traditional forms of media, such as television and talk radio, and social media) has amplified climate denial and misinformation. A special focus of inquiry might be Facebook's role in spreading climate misinformation, since several studies have highlighted its significant role. Students can also examine how algorithms create echo chambers and increase polarization; tap into and reinforce tendencies of motivated reasoning and cognitive biases; and generate potential for social harm when it further divides society, deepens political sectarianism, and makes concerted climate action based on reliable information more difficult.

In what ways should social media be regulated to ensure climate misinformation and climate denial aren't spread? This question requires students to wrestle with tensions between freedom of expression and its limits. They can examine the range of arguments related to social media regulation to prevent the spread of misinformation as well as questions about how or to what extent it might be regulated. Related questions include who should decide, what criteria might be used to make these decisions, what the roles and responsibilities of social media companies and cyber-citizens (rather than government) might be in managing these issues, and what might be the trade-offs between exercising free expression and limiting it in certain cases.

A related question is *Should media companies accept fossil fuel advertising dollars?* Given how important reliable, trustworthy sources are for all types of inquiries, investigation of what types of sources and media to trust merits ongoing attention, primarily because this can be complicated terrain to navigate. Consider two reputable and influential news sources: *The Guardian* and the *New York Times*. *The Guardian* has been an ardent supporter of the "keep it in the ground" initiative (a halt to all new fossil fuel development) and declared in 2020 that it will no longer accept advertising from fossil fuel extractive companies. While the *New York Times* has dedicated coverage to climate change and has reported or investigated climate denial, the company has continued to publish as well as produce "custom content" in its own studio for fossil fuel companies, including Shell, Chevron, and Exxon, by designing native advertising for these companies (AdsNotFitToPrint, n.d.).

The potential for algorithms, artificial intelligence, deep fakes, and doctored videos to mislead, misinform, and spread hard-to-detect falsehoods also offers opportunities for students to investigate the social harms and ethical issues related to these technologies and how they might be regulated in the public interest. Students can study how the biases of computer programmers, search engine algorithms, and data sets determine what they see in their online search results and social media feeds; how these biases can contribute to "filter bubbles" and science denial; and how the students and others might develop the algorithmic literacy necessary to understand how algorithms function and how they might be counteracted (Sinatra & Hofer, 2021).

The exploration of "ecomedia literacy" (Lopez, 2021) offers another set of inquiry possibilities. Students, for example, can investigate the ecological footprint of media use (range of costs tied to the production and distribution of computers, phones, and tablets along with immense electronic waste to manage) as well as its "ecological mindprint," including the impact of advertising on how people understand and respond to climate change (Lopez, 2021).

Geography

Where is climate denial a significant problem and why? Climate science is most politicized and polarized in the United States, England, and Australia (Painter, 2011). What accounts for this? Brüggemann and Engesser (2017) found that while the scientific consensus on global warming has grown more solid in recent decades, public opinion has remained uncertain about climate change in these three nations. Students might examine these nations as locations for major fossil fuel industry headquarters, the role of media in these countries, and how bitter partisan divide over climate change and political sectarianism in these countries contributes to this issue. Another site of significant climate denial is Brazil under President Jair Bolsonaro. Students might compare these different sites to identify factors across nations that contribute to denial.

What are the impacts of climate change on vulnerable populations in different parts of the world? Using ESRI (the Environmental Systems Research Institute) resources, students can engage in inquiries into climate migrants and refugees to highlight the impacts climate change is having on vulnerable populations in contexts like Alaska, the world's island nations, Darfur, Syria, and Bangladesh (see https://storymaps.esri.com/stories/2017/climate-migrants/).

How does the Global North–South divide inform explanations of climate change and denial? This question invites students to examine the locations of major carbon-emitting nations compared to the nations most affected by climate change, such as the Philippines, Mozambique, and the Bahamas (Eckstein et al., 2021). Studies of climate impacts on vulnerable populations in different locations of the world vividly demonstrate the effects of climate denial. Students can examine the practices of colonialism and imperialism and how they have contributed to global inequalities that affect climate change. Students can compare the imperial practices of major emitting nations like the United States and China, as well as conduct cross-national comparisons of the effects of climate change on countries in different parts of the world. A geography perspective also points to questions about climate reparations and the extent to which leading industrialized nations of the Global North should distribute funds to the Global South (and/or cancel debt) to address the Global North's historical contributions to climate change.

Of note, while we locate all the above questions within academic disciplines that align with broader contexts of denial, resources or tools from across the disciplines might be used to support an investigation. Consider a question like: *How is climate change denial part of my life, family, community?* This example of a personalized compelling question (Swan et al., 2018) could lead to investigations related to differences in consumption levels (e.g., homes, travel, fashion, food, and overall carbon emissions) among different groups, including the wealthiest elite compared to other groups within the United States and across the world. This inquiry question could also consider how educational, political, and business leaders or community organizers might be responding to the climate change threat—all of which point to ways concepts or tools from economics, politics, history, anthropology, sociology, psychology, and geography might be leveraged to address this question.

The Arts

As with just about any inquiry, the arts can be used to frame or facilitate investigations into subject matter. Visual art, theater or drama, music, and the digital arts, among other art-based modes or media, all offer a wide range of perspectives, ideas, and resources to enrich the inquiry pathways outlined above or help chart the exploration of other climate-related investigations. For example, students can examine how artists pursue different purposes—such as bearing witness to wildlife or species loss, glacier melt, or the forced dislocation of people and animals—raising awareness about climate change effects, or mobilizing action at a march or protest. More formal sites of education, including museums, can also be investigated with a climate change or denial lens. The Climate Museum, a nonprofit organization based in New York City, for example, integrates science education and the arts to help visitors grasp the climate crisis, which includes understanding how fossil fuel industry–led denial created the crisis. The museum also promotes equitable, just solutions (https://climatemuseum.org/mission) by showcasing the work of data journalist and artist Mona Chalabi, who created a series of posters for a project called *Beyond Lies* that targets corporate-led climate disinformation from the fossil fuel industry.

Art and media that use humor to raise awareness about climate change or motivate action to address its impacts can also support inquiry-based explorations. Some notable examples include satire about the absurdity of the "climate change debate" from the television host John Oliver (LastWeekTonight, 2014), the spoofing of oil company ads linked to Jimmy Kimmel (e.g., Saffer POV, 2021), as well as movies like *Don't Look Up* (McKay, 2021).

Of course, there is a long history of artists disrupting commonplace understandings and taken-for-granted assumptions, challenging deeply rooted stories-we-live-by. Many artists are provoking, evoking, or inspiring audiences to consider more sustainable and ecological visions of culture and society. For example, the *EcoArts Foundation* (https://www.ecoartsfoundation.org) brings together artists seeking to inspire cultural change and ecological regeneration through the arts. *Beautiful Rising* (https://beautifulrising.org) offers a range of resources for creative resistance across the Global South that include stories of campaigns and other actions; tactics for creative activists; and theories, principles, and methodologies to help grassroots movements use the arts to be more effective.

CONCLUSION

Ending this chapter with the arts is an ideal setup for the next and final chapter of the book, as we return to the *Climate Denial Inquiry Model* to develop the idea of *ecojustice stories-To-live-by*. Chapter 8 involves embracing new—or reclaiming lost—ways of seeing and being; the arts nurture and sustain these kinds of practices.

Ecojustice Stories-*To*-Live-By

> The climate crisis is, of course, only a symptom of a much larger crisis, a
> sustainability crisis, a social crisis, a crisis of inequality that dates back to
> colonialism and beyond, a crisis based on the idea that some people are
> worth more than others and therefore have the right to exploit and steal other
> people's lands and resources. And it is very naive to believe that we can solve
> this crisis without confronting the roots of it.
>
> —Greta Thunberg (October 2021)

To some extent, how, or whether or not, to confront the "roots" of the crisis highlights key distinctions among climate action leaders. For some, like climate scientist Michael Mann (2021), capsizing the existing economic system of capitalism, for example, is ill-advised; system change might be a longer-term solution, but a host of actions can and need to be taken now to avert more severe climate impacts. For others, such as Naomi Klein (2014), humanity is presented with a stark choice; it's either capitalism or the climate because the reality of the climate crisis "changes everything." With this view, more radical reforms are necessary, such as "making extreme wealth extinct" (Monbiot, 2021).

While this book is not focused on the exploration and evaluation of climate-related solutions, our starting point is to understand climate change in terms of (in)justice. Adverse effects of climate change are felt disproportionately around the world with poorer countries and communities far more likely to experience climate-related food and water insecurity, dislocation, and forced migration. A historical perspective also highlights extensive connections between climate change and economics, race, gender, and colonization. The co-chair of the Climate Justice Alliance, Elizabeth Yeampierre, draws a "direct line from slavery and rapacious exploitation of natural resources to current issues of environmental justice" (Gardiner, 2020). Yeampierre points out that climate change stems from a legacy of colonialism, slavery, and land and resource extraction. She notes:

> A lot of times when people talk about environmental justice they go back to the
> 1970s or '60s. But I think about the slave quarters. I think about people who got

the worst food, the worst health care, the worst treatment, and then when freed, were given lands that were eventually surrounded by things like petrochemical industries. The idea of killing black people or indigenous people, all of that has a long, long history that is centered on capitalism and the extraction of our land and our labor in this country. (In Gardiner, 2020)

Bringing this history into the present, Reverend Lennox Yearwood Jr., (2020), a minister and community activist and president of the Hip Hop Caucus, a national nonprofit, nonpartisan organization committed to youth political empowerment, contends that the climate crisis and environmental injustice are bound together because they "play out within the same systems of white supremacy and structural racism" (para. 8).

Tuck and Yang (2012) outline the need to unsettle ignorance and innocence: The idea of reconciliation with systems of settler colonialism, white supremacy, and structural racism based on the theft and destruction of Indigenous lands is untenable. Speaking about "hope in common," Graeber (2008) notes that we have created a "vast bureaucratic apparatus for the creation and maintenance of hopelessness, a giant machine, designed, first and foremost, to destroy any sense of possible alternative futures" (para. 3). Graeber suggests there is a need to engage with possible alternative futures:

> What remains is what we are able to promise one another. Directly. Without the mediation of economic and political bureaucracies. The revolution begins by asking: what sort of promises do free men and women make to one another, and how, by making them, do we begin to make another world? (final para.)

Given the realities of our historical moment, it is time to contest intricately crafted, complex, and destructive stories-we-*have*-lived-by and "restory" our relationships with each other and the world (Damico et al., 2020; Kimmerer, 2013; Latour, 2018). One compelling way to do this is by investigating *ecojustice stories-To-live-by*.

THREE ECOJUSTICE STORIES-*TO*-LIVE-BY

Grounded in ideas of Earth democracy (Shiva, 2005) that emphasize cultural and natural diversities, human and ecological well-being, and concern for social and ecological justice (Peters, 2017), we explore three interrelated *stories-To-live-by*. While not necessarily "new," and by no means exhaustive, these stories align with cultural traditions that can help address our current crises.

1. All Life Is Treated With Respect, Care, and Responsibility, Especially Our Most Vulnerable Populations and Species

This *ecojustice story-To-live-by* is grounded in a feminist ethic of care, concern, and connection (Martin, 1994). Capacities for care require challenging hierarchies and forms of domination to ensure there is equal access to the means of reproduction that are the foundation of all life (Federici, 2018). Care and responsibility entail the work of listening, cooperating, discussing, negotiating, and learning to deal with disagreement to build a sense of community. It is a process of shifting power dynamics in favor of those who are "left behind," marginalized, and most vulnerable. It views caregiving (health care, child care, elder care, care for one's community, care for the environment, and care for the most vulnerable) and the collective reproduction of life as among the highest purposes of individuals and society.

The work of the Mary Robinson Foundation–Climate Justice (https://www.mrfcj.org/) aligns with this *ecojustice story-to-live-by* by showcasing how key principles of climate justice, such as gender equality and equity, sharing benefits and burdens equitably, protecting human rights, building effective partnerships to secure climate justice, and harnessing education for climate stewardship are being enacted across the world. The podcast series Mothers of Invention (https://www.mothersofinvention.online/), for example, highlights intergenerational activists whose climate justice gender work intersects with issues related to race, colonialism, social class, and poverty. The podcast offers powerful models of women working for climate justice, such as the Black Lives Matter activist Sarra Tekola, the daughter of a climate refugee from Ethiopia. Tekola calls for community-based solutions to climate change. She connects feminism and climate justice with the Black Lives Matter movement to help people understand and address issues of state violence toward women of color as deeply rooted in particular colonialist legacies. Tekola is a founding member of Women of Color Speak Out (https://wocspeakout.com/), a collective of Seattle activists working to educate their community about the climate crisis by connecting climate action to other social movements that help people see that it is possible to dismantle deeply rooted forms of oppression and imagine more sustainable futures. Of note, this story about respect, care, and responsibility is less about altering who we are as humans and more about reclaiming that "our extraordinary capacity for altruism and our remarkably social nature are the central, crucial facts about humankind" (Monbiot, 2017).

2. The Primary Goal of Society Is Human, Ecological, and Planetary Well-Being That Comes With a Recognition of Limits

"The care of human life and happiness, and not their destruction, is the first and only task of good government," wrote Thomas Jefferson in 1809. People have a right to pursue healthy and meaningful lives, rewarding and purposeful work, and opportunities to develop their full human capabilities (Nussbaum, 2011; Sen, 1999). Jackson (2021) suggests this is "a story that recognizes the breadth and depth of the human soul . . . that accounts for affinity with tradition as well as our desire for novelty . . . that offers us the tantalizing freedom to become not less, but more fully human" (p. 98). All of these facets of being human are core to flourishing, along with the recognition that humans can only flourish if our natural surroundings and all of life flourishes. The goal of society and government, then, is to ensure life (in all its forms) can flourish. This includes defending the natural environment, resisting the theft, occupation, and destruction of land in order to restore and protect it (Sabzalian, 2019).

The well-being of our ecosystems requires a recognition of limits that are necessary for more-sustainable visions of life. Science-based analyses of the Earth's environmental limits provide a framework to "define a safe operating space for human societies to develop and thrive" (Steffen et al., 2015, p. 737). Ecological limits also require recognition of ethical and political limits—the need for individuals and societies to learn the self-restraint necessary to control desires and impulses as well as the limits to the influence of private interests and corporate power necessary in a well-functioning democracy (Baildon & Damico, 2019).

The perspectives, practices, and traditions of Indigenous peoples are central to this *ecojustice story*. For example, the First Nations Health Authority in British Columbia highlights a holistic vision of health and wellness based on cultural wisdom, respect, responsibility, and relationships that cut across environmental, social, cultural, and economic determinants of well-being grounded in self-determining communities (https://www.fnha.ca/). In this vision, human and ecological well-being is based on recognition of limits and the need for community action to defend and restore land and culture against the destructive stories that people are separate from nature and that the primary goal of society is limitless economic growth. Indigenous well-being is grounded in narratives of place that focus less on rights (to private property, to extract resources and destroy lands, etc.) and more on "individual and collective responsibilities, authority, and obligations, and how that changes based on who you are, where you are, and what you are doing" (Aikau et al., 2015, p. 87). These narratives connect place, one's sense of identity, and the responsibilities necessary to exercise freedom in ways that recognize our interconnectedness with place and all living things (Hester, 2010).

Our capacity to imagine and enact new stories-To-live-by and different forms of social arrangement within ecological limits is part of what makes us human (Graeber & Wengrow, 2021). People are creating alternative visions of the future that are more just and sustainable. These include postgrowth (Jackson, 2021; Prádanos, 2018), postmaterialist (Inglehart, 1977), and postcapitalist (Mason, 2015) imaginaries that challenge dominant growth paradigms of GDP, profit, productivity, and progress. The futurist Hazel

Henderson has advocated ethical markets and research-based indicators that challenge GDP and other traditional economic metrics as measures of progress and well-being. With the Calvert Group, Henderson developed the Calvert-Henderson Quality of Life Indicators that consider twelve domains of life: education, employment, energy, environment, health, human rights, income, infrastructure, national security, public safety, recreation, and shelter (Henderson, et al, 2000). These indicators recognize that human, ecological, and planetary well-being requires new ways of thinking about economics, progress, and our collective future.

Henderson is also the founder of Ethical Markets (ethicalmarkets.com) and has developed a Green Transition Scoreboard to track green investments in renewable energy, energy efficiency, life systems (water, waste, recycling, etc.), green construction, and green corporate research and development to argue that "the green economy [is] growing faster than anyone realizes" (https://www.ethicalmarkets.com/2017-green-transition-scoreboard-tracks -private-green-investments-at-8-1-trillion/). Just as the fossil fuel divestment movement is gaining in strength, activist shareholders in U.S. banks are putting forth proposals to stop the financing of new fossil fuel projects (Andreoni, 2022). Along with the efforts in many communities to use renewable energies, electrify transportation, remove carbon, and protect nature, there are hopeful signs that climate denial can be overcome by creating alternative ways of living and organizing our societies. The Postgrowth Encyclopedia (https://www .postgrowth.org/the-post-growth-encyclopedia) offers concepts and principles (e.g., buy local, ethical consumption, co-housing, passive solar design), models and indicators (e.g., voluntary simplicity, Gross National Happiness, Genuine Progress Indicators), and activities, programs, and movements (e.g., cycling, transition towns, sociocracy, collaborative consumption) that spotlight the many ways people are reimagining and revitalizing their communities to promote collective well-being within ecological limits.

3. Civic Engagement for the Common Good Is Necessary for More Just and Meaningful Lives and Futures

In contrast to unfettered valorization of individual freedoms and upholding the virtues of competition and a zero-sum story, this *ecojustice story-To-live-by* attaches importance and deep value to collective efforts to improve people's lives. Again, Indigenous groups have continuously resisted the occupation and destruction of their lands and erasures of culture and sovereignty by settler colonialism, while calling for more holistic and inclusive forms of justice, social and ecological responsibility, and place-based political action (Sabzalian, 2019). Indigenous-led actions and movements led by groups like Idle No More and Indigenous Climate Action highlight this ecojustice story in action. The Indigenous Climate Action website (https://www .indigenousclimateaction.com/), for example, provides resources and tools to ensure Indigenous knowledge and skills serve as drivers of climate justice and brings people together to build relationships and deliberate climate action. The System Change Not Climate Change Project (https://canadians.org /systemchange) offers multimedia tools for climate justice organized by the Council of Canadians' Climate Justice for People and the Planet campaign. The website stemming from the *This Changes Everything* documentary film (Lewis et al., 2015) and book (Klein, 2014) narrates an ecojustice story with a study guide and lesson plans, including a lesson plan called "Reinventing a Clean and Just Economy" (https://thischangeseverything.org/wp-content /uploads/2015/12/Lesson-Plan-5.pdf), which explores the story of Henry Red Cloud, who left his steel industry job to start his own solar-power company and create jobs for those living in native communities. In this example, students can learn about how efforts like this yield a synergy of benefits for communities, the economy, and the environment, as Henry aims to help First Nations people achieve energy independence. Similarly, biologist, lyrical science writer, and member of the Potawatomi Nation Robin Wall Kimmerer (2013) writes about the braiding together of healing stories from science, literature, and Indigenous knowledge to imagine and create a different relationship with the world, one based on regenerative, reciprocal relationships.

Civic engagement for the common good is also about revitalizing "the commons," the natural and cultural systems that must be noncommercialized and nonprivatized—the air, water, plants, animals, woodlands, and forms of knowledge (narratives, aesthetic and craft knowledge, moral norms and traditions, etc.) that are life-sustaining (Bowers, 2004). Federici (2018) calls for a politics of the commons as a collective struggle against the ways people have been divided from each other and the land and for a more democratic vision of the common good and of the future.

This ecojustice story-to-live-by is deeply rooted in youth-led climate activism and civic engagement. Projects such as Zero Hour (http://thisiszerohour .org/), the Sunrise Movement (https://www.sunrisemovement.org/), Extinction Rebellion (https://rebellion.earth/), Fridays for Future (https://fridaysforfuture .org/), and School Strike for Climate (https://www.schoolstrike4climate.com/) are grassroots efforts to shape climate-related public policy, grounded deeply in what longtime consumer advocate and public intellectual Ralph Nader (2019) calls the moral power of youth. Extinction Rebellion, for example, is an international network dedicated to nonviolent direct action to convince governments to respond swiftly to our climate and ecological emergency. The Sunrise Movement in the United States has also been highly engaged in political action, leading efforts to create a Green New Deal that will confront climate change and economic inequality.

There are also many other groups, organizations, and individuals whose efforts embody an ecojustice story about civic engagement for the common good being an essential part of living a purposeful life. This list includes organizations fighting the climate crisis and promoting social and racial justice, like

350.org, Friends of the Earth, the Indigenous Environmental Network, and Greenpeace, among many others (the Climate Action Network includes more than 1,500 civil organizations in 130 countries); global, multilateral associations like the Intergovernmental Panel on Climate Change, which has 120 member nations; and the many scientific and citizen communities, institutions, and think tanks across a range of fields (environmentalism, law, human rights, animal welfare, among others) all advocating for or engaging in climate action (Hestres, 2020). One prominent political force has been the fossil fuel divestment movement.

The life's work of Ralph Nader, tireless consumer advocate, environmentalist, political activist, and public intellectual, also embodies the essence of this *ecojustice story-to-live-by*. His laser-beam focus on the abuses of corporate power with clear, straightforward ideas about how individuals and groups can "break through power" to revitalize (and re-democratize) communities is readily accessible across his many books and talks (e.g., Nader, 2016b), and the Ralph Nader Radio Hour podcast, along with resources for youth that chronicle his role in fighting for our rights as citizens and consumers (Panchyk, 2021).

The law, and using the courts to advance climate justice, remains a particularly robust way to put this third *ecojustice story-to-live-by* into practice. Dozens of lawsuits have targeted major fossil fuel giants like BP, ExxonMobil, and Chevron, charging these companies with deceiving the public and exacerbating losses to states and cities or asking that these companies pay affected communities for climate-related costs (Klein & Stefoff, 2021; Sokol, 2020). The United States Government has even been the target of a high-profile lawsuit. In 2015, in *Juliana v. United States*, 21 youth plaintiffs sued the executive branch for policies that have encouraged or supported more fossil fuel extractivism and accelerated climate change, which knowingly violated their rights of life, liberty, and property. There are also efforts from groups like Law Students for Climate Accountability that target the law firms that have worked in service of the fossil fuel industry.

FROM SOCIAL ORGANIZATION OF DENIAL TO COLLECTIVE COORDINATION FOR ECOJUSTICE

In her study of a rural Norwegian community during an unusually warm winter, Kari Norgaard (2011) shows how the community did not integrate their climate knowledge into their lives; instead, people opted to distance themselves from responsibility as they asserted the "rightness or goodness" of their actions and maintained "order and security" as they constructed "a sense of innocence in the face of the disturbing emotions associated with climate change" (pp. 11–12). Given the social organization of denial (Zerubavel, 1997, 2002) in this community, climate realities collided with the story of the community's "goodness" and innocence and its desire to maintain the status quo.

In the United States the social organization of climate denial has been alive and well in education across our professional organizations, national, state- and local-level associations, schools of education in universities, individual school districts, and in our classrooms. More specific to this book's purposes, there has not been a comprehensive reckoning with the climate crisis in literacy and social studies education. In many ways, "our heads have been in the sand too long" (Amigo Fields, 2021).

Yet in May 2022 there is less outright denial of climate science in the United States along with greater recognition that more needs to be done to address the climate crisis. As educators committed to teaching, researching, and learning about climate change dating back to 2008, we are heartened by this development. It is clearly a step in the right direction. Many big and bold steps still need to be taken. Climate denial remains pervasive and protean as climate denial purveyors continue to produce an ever-changing array of texts or stories that inadequately address the climate crisis.

This is why the two of us side with climate scientists like Michael Mann (2021) and Kimberly Nicholas (2021) who continue to assert that while

It's warming
It's us
We're sure
It's bad
We can fix it.

In this book we have made a case that "fixing it" entails a straightforward confrontation with climate denial and dislodging destructive stories that we have been living by. Making climate denial a curricular and instructional priority through inquiry-based investigations can lead to pathways toward *ecojustice stories-To-live-by*. This is one way we, as educators—teachers, administrators, parents, and community members alike—can all do our part in helping "fix it."

Of course, there is no definitive "fix" to climate denial. Climate denial is pervasive and protean as its purveyors continue to promulgate an array of techniques, texts, or stories that inadequately address the climate crisis. This is why each of us will need to play our part and in so doing, know we will not be alone. The three *ecojustice stories-To-live-by* are fundamentally about belonging; each speaks to a deeper recognition that rather than walking, working, and feeling alone, we can "restore our sense of belonging: belonging to ourselves, belonging to our communities, belonging to our localities, belonging to the world" (Monbiot, 2017). As we collectively confront climate denial and advance ecojustice, we can embrace a fundamental truth: We all need each other (McGhee, 2021). This might be the most elemental truth of all.

Appendix

Throughout this book we have highlighted the critically important and inspiring work of many people and projects. This appendix highlights a few web-based resources for understanding climate denial and how to teach about it. Rather than present a more exhaustive list here, we limited ourselves to identifying 10 resources we think can serve as a sturdy foundation and manageable starting point to ground much further exploration and investigation.

UNDERSTANDING CLIMATE DENIAL

DeSmog: https://www.desmog.com/
A global organization dedicated since 2006 "to clear the PR pollution that is clouding the science and solutions to climate change." DeSmog provides consistent and thorough investigative reporting.

Drilled: https://drillednews.com/podcast-2/
This "true-crime podcast about climate change" is hosted by Amy Westervelt. We anxiously await the release of each new episode. Just a critical resource.

Inside Climate News: https://insideclimatenews.org/
Pulitzer Prize–winning, nonpartisan reporting on the biggest crisis facing our planet. A vital source of investigative reporting.

Covering Climate Now: https://coveringclimatenow.org/
This organization was created by journalists to support other journalists and newsrooms in producing informed, compelling, and urgent climate stories. The primary goal is collaboration of news outlets across the globe to cover climate change.

Climate Feedback: https://climatefeedback.org/
A global network of scientists that analyzes climate change media coverage to discern trustworthy information.

TEACHING ABOUT CLIMATE DENIAL

Action for the Climate Emergency (formerly Alliance for Climate Education): https://acespace.org/
    A nonprofit organization that offers educational resources about climate science and climate justice along with leadership training for youth.

Brown University. *Climate Change and Questions of Justice.* https://www.choices.edu/curriculum-unit/climate-change-questions-justice/

Klein, N. This Changes Everything: Study guide and lesson plans. https://thischangeseverything.org/studyguide/

New York Times. Resources for Teaching About Climate Change with the *New York Times.* https://www.nytimes.com/2021/11/04/learning/lesson-plans/resources-for-teaching-about-climate-change-with-the-new-york-times.html

Rethinking Schools & Teaching for Change. (2019). *Zinn education project.* https://www.zinnedproject.org/ (search for "climate justice").

References

Abdul-Razzak, N., Prato, C., & Wolton, S. (2019, March 1). After Citizens United: How outside spending shapes American democracy. *SSRN*. http://dx.doi.org/10.2139/ssrn.2823778

AdsNotFitToPrint. (n.d.). *Tell* The New York Times: *Stop promoting fossil fuels!* https://adsnotfittoprint.com/

Aikau, H. K., Arvin, M., Goeman, M., & Morgensen, S. (2015). Indigenous feminisms roundtable. *Frontiers: A Journal of Women Studies, 36*(3), 84–106.

Albrecht, G. A. (2019). *Earth emotions: New words for a new world*. Cornell University Press.

Allen, D. (2004). *Talking to strangers: Anxieties of citizenship since* Brown v. Board of Education. University of Chicago Press.

Amigo Fields. (2021). Meet us in the street [Song]. https://amigofields.hearnow.com/

Anang, M. (2021). *Law firm climate change scorecard*. Law Students for Climate Accountability. https://www.ls4ca.org/

Andreoni, M. (2022, May 10). How shareholders are pushing big banks for climate action. *The New York Times*. https://www.nytimes.com/2022/05/10/climate/banks-shareholders-climate-action.html

Andreotti, V. (2006). Soft versus critical global citizenship education. *Development Education, Policy and Practice, 3*(1), 40–51.

Anker, E. R. (2021). *Ugly freedoms*. Duke University Press.

Appiah, A. (2018). *The lies that bind: Rethinking identity, creed, country, color, class, culture*. Liveright.

Archer, M. S. (2012). *The reflexive imperative in late modernity*. Cambridge University Press.

Arendt, H. (1958). *The human condition*. University of Chicago Press.

Arendt, H. (1977). *Between past and future: Eight exercises in political thought*. Penguin Books.

Aronczyk, M., & Espinoza, M. I. (2021). *A strategic nature: Public relations and the politics of American environmentalism*. Oxford University Press.

Aronoff, K. (2020, February 19). Obama's climate legacy and the lie of "energy independence." *The New Republic*. https://newrepublic.com/article/156580/obamas-climate-legacy-lie-energy-independence

Aronoff, K. (2021). *Overheated: How capitalism broke the planet – and how we fight back*. Bold Type Books.

Avaaz. (2021, May 11). *Facebook's climate of deception: How viral misinformation fuels the climate emergency*. https://avaazimages.avaaz.org/facebook_climate_misinformation.pdf

Baildon, M., & Damico, J. S. (2011a). *Social studies as new literacies in a global society: Relational cosmopolitanism in the classroom*. Routledge/Taylor & Francis.

Baildon, M., & Damico, J. S. (2011b). Judging the credibility of Internet sources: Developing critical and reflexive readers of complex digital texts. *Social Education, 75(5)*, pp. 269–273.

Baildon, M., & Damico, J. S. (2019). Education in an age of limits. *Journal of Curriculum Theorizing, 34(3)*, 25–40. https://journal.jctonline.org/index.php/jct/article/view/823

Banerjee, N., Gustin, G., & Cushman, J. H. (2018, December 21). *The Farm Bureau: Big oil's unnoticed ally fighting climate science and policy*. Inside Climate News. https://insideclimatenews.org/news/21122018/american-farm-bureau-fossil-fuel-nexus-climate-change-denial-science-agriculture-carbon-policy-opposition/

Barrett, L. (2017). *How emotions are made: The secret life of the brain*. Houghton Mifflin Harcourt.

Bartels, L. M. (2009). Economic inequality and political representation. In L. Jacobs & D. King (Eds.), *The unsustainable American state* (pp. 167–196). Oxford University Press.

Barton, K. C. (2005). Primary sources in history: Breaking through the myths. *Phi Delta Kappan, 86(10)*, 745–753.

Barton, K. C., & Ho, L. (2021). *Curriculum for justice and harmony*. Routledge.

Barton, K. C., & McCully, A. (2007). Teaching controversial issues . . . where controversial issues really matter. *Teaching History, 127*, 13–19.

Beach, R., Share, J., & Webb, A. (2017). *Teaching climate change to adolescents: Reading, writing, and making a difference*. Routledge.

Beck, I. L., McKeown, M. G., & Worthy, J. (1995). Giving a text voice can improve students' understanding. *Reading Research Quarterly, 30(2)*, 220–238.

Bigelow, B. (2014). How my schooling taught me contempt for the earth. In B. Bigelow & T. Swinehart (Eds.), *A people's curriculum for the earth: Teaching climate change and the environmental crisis* (pp. 36–41). Rethinking Schools Limited.

Bigelow, B., & Swinehart, T. (Eds.). (2014). *A people's curriculum for the earth: Teaching climate change and the environmental crisis*. Rethinking Schools Limited.

Bjornberg, K. E., Karlsson, M., Gilek, M., & Hansson, S. O. (2017). Climate and environmental science denial: A review of the scientific literature published in 1990–2015. *Journal of Cleaner Production, 167*, 229–241.

Blumenthal, D. (2019, February 3). The unpredictable rise of China. *The Atlantic*. https://www.theatlantic.com/ideas/archive/2019/02/how-americans-misunderstand-chinas-ambitions/581869/

Bó, E. D. (2006). Regulatory capture: A review. *Oxford Review of Economic Policy, 22(2)*, 203–225.

Bodor, Á., Varjú, V., & Grünhut, Z. (2020). The effect of trust on the various dimensions of climate change attitudes. *Sustainability* 2020, *12(23)*, 10200. https://doi.org/10.3390/su122310200

Boler, M. (1999). *Feeling power: Emotions and education*. Routledge.

Boykoff, M., & Farrell, J. (2020). Climate change countermovement organizations and media attention in the United States. In N. Almiron & J. Xifra (Eds.), *Climate change denial and public relations* (pp. 121–139). Routledge.

Boren, Z., Kaufmann, A. C., & Carter, L. (2020, September 28). Revealed: BP and Shell back anti-climate lobby groups despite pledges. *Huffington Post*. https://www.huffpost.com/entry/bp-shell-climate_n_5f6e3120c5b64deddeed6762

Bowers, C. A. (1997). *The culture of denial: Why the environmental movement needs a strategy for reforming universities and public schools*. State University of New York Press.

Bowers, C. A. (2001). *Educating for eco-justice and community*. University of Georgia Press.

Bowers, C. A. (2004). Revitalizing the commons or an individualized approach to planetary citizenship: The choice before us. *Educational Studies, 36*(1). https://doi.org/10.1207/s15326993es3601_5

Bowers, C. A. (2006). *Revitalizing the commons: Cultural and educational sites of resistance and affirmation*. Lexington Books.

Bowers, C. A. (2017). Educational reforms for survival. *Tikkun, 32*(4), 25–32.

Branch, G. (2020, June 24). Op-ed: Follow-up needed on NJ's renewed commitment to climate change education. *NJ Spotlight News*. https://www.njspotlight.com/2020/06/op-ed-follow-up-needed-on-njs-renewed-commitment-to-climate-change-education/

Branch, G., Rosenau, J., & Berbeco, M. (2016). Climate education in the classroom: Cloudy with a chance of confusion. *Bulletin of the Atomic Scientists, 72*(2), 89–96. http://dx.doi.org/10.1080/00963402.2016.1145906

Bressler, R. D. (2021). The mortality cost of carbon. *Nature Communications, 12*(1). https://doi.org/10.1038/s41467-021-24487-w

Britzman, D. (1998). *Lost subjects, contested objects. Toward a psychoanalytic inquiry of learning*. State University of New York Press.

Brüggemann, M., & Engesser, S. (2017). Beyond false balance: How interpretive journalism shapes media coverage of climate change. *Global Environmental Change, 42*(2017), 58–67.

Brulle, R. J. (2014). Institutionalizing delay: Foundation funding and the creation of U.S. climate change counter-movement organizations. *Climatic Change, 122*(4), 681–694. https://doi.org/10.1007/s10584-013-1018-7

Brulle, R. J., Aronczyk, M., & Carmichael, J. (2019). Corporate promotion and climate change: An analysis of key variables affecting advertising spending by major oil corporations, 1986–2015. *Climatic Change, 159*, 87–101. https://doi.org/10.1007/s10584-019-02582-8

Brulle, R. J., & Werthman, C. (2021). The role of public relations firms in climate change politics. *Climatic Change, 169*(1), 1–21.

Buchan, S. (2021, May 5). *On the back burner: How Facebook's inaction on misinformation fuels the global climate crisis*. Stop Funding Heat. https://stopfundingheat.info/wp-content/uploads/2021/05/On-The-Back-Burner-Final.pdf

Bullard, R. D. (2020, October 2). *Building a just, fair and equitable society in the era of climate change*. [Keynote address]. Midwest Climate Summit. https://midwestclimatesummit.wustl.edu/event-details-2/past-sessions-recordings/

Burbules, N. C. (1993). *Dialogue in teaching: Theory and practice*. Teachers College Press.

Busch, K. C., & Osborne, J. (2014). Effective strategies for talking about climate change in the classroom. *School Science Review, 96*(354), 25–32.

Carrington, D. (2021a, April 19). "A great deception": Oil giants taken to task over "greenwash" ads. *The Guardian*. https://www.theguardian.com/business/2021/apr/19/a-great-deception-oil-giants-taken-to-task-over-greenwash-ads

Carrington, D. (2021b, October 6). Fossil fuel industry gets subsidies of $11m a minute, IMF finds. *The Guardian*. https://www.theguardian.com/environment/2021/oct/06/fossil-fuel-industry-subsidies-of-11m-dollars-a-minute-imf-finds

Cassidy, J. (2020, February 3). Can we have prosperity without growth? *The New Yorker.* https://www.newyorker.com/magazine/2020/02/10/can-we-have-prosperity-without-growth

Center for Humane Technology. (2021). *Ledger of harms.* https://ledger.humanetech.com/

Center for International Environmental Law (CIEL). (2019, December 9). *Groundbreaking inquiry in Philippines links carbon majors to human rights impacts of climate change, calls for greater accountability.* https://www.ciel.org/news/groundbreaking-inquiry-in-philippines-links-carbon-majors-to-human-rights-impacts-of-climate-change-calls-for-greater-accountability/

Chakraborty, J., Collins, T. W., & Grineski, S. E. (2019). Exploring the environmental justice implications of Hurricane Harvey flooding in Greater Houston, Texas. *American Journal of Public Health, 109*(2), 244–250.

Chen, G. (2018, August 28). *States fight to teach "ignorant" science to students.* Public School Review. https://www.publicschoolreview.com/blog/states-fight-to-teach-ignorant-science-to-students

Chesterman, S., Malone, D., Villalpando, S., & Ivanovic, A. (2019). *The Oxford handbook of United Nations treaties.* Oxford University Press.

Chuvakhina, L. G., Moldenhauer, N. A., & Nasirbeik, A. (2021). Assessment of the impact of US energy policy on world oil prices. *Journal of Mines, Metals and Fuels, 69*(4), 111–119. https://doi.org/10.18311/jmmf/2021/27798

Climate Change Commission. (2018). *Climate change and the Philippines* [Executive Brief No. 2018-1]. Philippines Climate Change Commission. https://niccdies.climate.gov.ph/climate-change-impacts

Coan, T., Boussalis, C., Cook, J., & Nanko, M. O. (2021, March 9). *Computer-assisted detection and classification of misinformation about climate change.* SocArXiv Papers. https://doi.org/10.31235/osf.io/crxfm

Cohen, S. (2001). *States of denial: Knowing about atrocities and suffering.* Polity.

Collins, C. (2021, October 18). *Updates: Billionaire wealth, U.S. job losses and pandemic profiteers.* Inequality.org. https://inequality.org/great-divide/updates-billionaire-pandemic/

Comber, B. (2013). Critical literacy in the early years: Emergence and sustenance in an age of accountability. In J. Larson & J. Marsh (Eds.), *The SAGE handbook of early childhood literacy* (pp. 587–601). SAGE.

Cook, J. (2020a). *Cranky uncle vs. climate change: How to understand and respond to climate deniers.* Citadel Press Books.

Cook, J. (2020b). Deconstructing climate science denial. In D. Holmes & L. M. Richardson (Eds.), *Edward Elgar research handbook in communicating climate change* (pp. 62–78). Edward Elgar.

Cook, J., & Jacobs, P. (2017). FLICC: the five traits of science denial. https://www.youtube.com/watch?v=WgsjiWpptHw

Cook, J., Nuccitelli, D., Green, S. A., Richardson, M., Winkler, B., Painting, R., Way, R., Jacobs, P., & Skuce, A. (2013). Quantifying the consensus on anthropogenic global warming in the scientific literature. *Environmental Research Letters, 8*(2), 1–7. http://stacks.iop.org/1748-9326/8/i=2/a=024024

Cooke, N. A. (2017). Posttruth, truthiness, and alternative facts: Information behavior and critical information consumption for a new age. *The Library Quarterly, 87*(3), 211–221.

Cooper, B. (2021, March 11). *Rosneft, Gazprom, and Russia's failure to adopt green policies*. Foreign Policy Research Institute. https://www.fpri.org/article/2021/03/rosneft-gazprom-and-russias-failure-to-adopt-green-policies/

Corruption at the heart of the United Nations. *The Economist*. (2005, August 9). https://www.economist.com/unknown/2005/08/09/corruption-at-the-heart-of-the-united-nations

D'Alisa, G., Demaria, F., & Kallis, G. (Eds.). (2015). *Degrowth: A vocabulary for a new age*. Routledge.

Damico, J. S. (2021). Then and now: Time for ecojustice literacies. *Language Arts, 99*(2).

Damico, J. S., & Baildon, M. (2011). Content literacy for the 21st century: Excavation and elevation and relational cosmopolitanism in the classroom. *Journal of Adolescent and Adult Literacy, 55*(3), 232–243.

Damico, J. S., & Baildon, M. (2015). Rethinking reliability after students evaluate a Facebook page about health care in Singapore. *Journal of International Social Studies, 5*(1), 51–63. Available at: http://www.iajiss.org/index.php/iajiss/article/view/156/169

Damico, J., Baildon, M., Exter, M., & Guo, S.-J. (2009). Where we read from matters: Disciplinary literacy in a 9th grade social studies classroom. *Journal of Adolescent & Adult Literacy, 5*(4), 325–335.

Damico, J. S., Baildon, M., & Panos, A. (2018). Media literacy and climate change in a post-truth society. *Journal of Media Literacy Education, 10*(2), 11–32. https://digitalcommons.uri.edu/jmle/vol10/iss2/2/

Damico, J. S., Baildon, M., & Panos, A. (2020). Climate justice literacy: Stories-we-live-by, ecolinguistics, and classroom practice. *Journal of Adolescent & Adult Literacy, 63*(6), 683–691. https://doi.org/10.1002/jaal.1051

Damico, J. S., Campano, G., & Harste, J. (2009). Transactional and critical theory and reading comprehension. In S. Israel and G. Duffy (Eds.), *Handbook of research on reading comprehension* (pp. 177–188). Lawrence Erlbaum.

Damico, J. S., Honeyford, M., & Panos, A. (2016). Designing inquiries that matter: Targeting significance, diversity, and fit. *Voices in the Middle, 23*(3), 25–32.

Damico, J. S., & Panos, A. (2016). Reading for reliability: Preservice teachers evaluate web sources about climate change. *Journal of Adolescent & Adult Literacy, 60*(3), 275–285.

Damico, J. S., & Panos, A. (2018). Civic Literacy as 21st century source work: Future social studies teachers examine web sources about climate change. *Journal of Social Studies Research, 42*(4), 345–359. https://www.sciencedirect.com/science/article/pii/S0885985X17301365

Damico, J. S., Panos, A., & Baildon, M. (2018). "I'm not in the truth business": The politics of climate change with pre-service teachers. *English Teaching: Practice & Critique. 17*(2), 72–89.

Damico, J. S., Panos, A., & Myers, M. (2018). Digital literacies and climate change: Exploring reliability and truth(s) with pre-service teachers. In E. Ortlieb, E. Cheek Jr., & P. Semingson (Eds.), *Best practices in teaching digital literacies* (pp. 93–108).

Davis, C. (2006, May 6). Research at MU: News and Press Releases. https://archive.thinkprogress.org/climate-scientist-to-cei-stop-misrepresenting-my-research-b886fd5cd5e1/

DeSmog. (n.d.). Committee for a Constructive Tomorrow (CFACT). https://www
.desmog.com/committee-constructive-tomorrow/

Dewey, J. (1916). *Democracy and education: An introduction to the philosophy of education.* Macmillan.

Diele-Viegas, L. M., & Rocha, C.F.D. (2020). Why releasing mining on Amazonian indigenous lands and the advance of agrobusiness is extremely harmful for the mitigation of world's climate change? Comment on Pereira et al. (*Environmental Science & Policy,* 100(2019), 8–12). *Environmental Science & Policy, 103*(2020), 30–31.

Doerr, J. (2021). *Speed and scale: A global action plan for solving our climate crisis now.* Penguin.

Drennen, A., & Hardin, S. (2021, March 30). *Climate deniers in the 117th Congress.* Center for American Progress. https://www.americanprogress.org/article/climate -deniers-117th-congress/

Druckman, J. N., & McGrath, M. C. (2019). The evidence for motivated reasoning in climate change preference formation. *Nature Climate Change, 9*(2), 111–119.

Drutman, L. (2020). Breaking the two-party doom loop: The case for multiparty democracy in America. *Oxford Scholarship Online.* doi:10.1093/oso/97801909 13854.001.0001.

Drutman, L. (2021, September 8). Quiz: If America had six parties, which would you belong to? *The New York Times.* https://www.nytimes.com/interactive/2021/09 /08/opinion/republicans-democrats-parties.html

Dryzek, J. S., Norgaard, R. B., & Schlosberg, D. (2011). Climate change and society: Approaches and responses. In J. S. Dryzek, R. B. Norgaard, & D. Schlosberg (Eds.), *The Oxford handbook of climate change and society.* Oxford University Press. doi:10.1093/oxfordhb/9780199566600.003.0001

Duggan-Haas, D., & Ross, R. M. (2017). *The teacher-friendly guide to climate change.* Paleontological Research Institution.

Dunlap, R. E. (2008). The new environmental paradigm scale: From marginality to worldwide use. *Environmental Education, 40*(1), 3–10.

Dunlap, R. E., & McCright, A. M. (2010). Climate change denial: Sources, actors and strategies. In C. Lever-Tracy (Ed.), *Routledge handbook of climate change and society* (pp. 270–290). Routledge.

Dunlap, R. E., & McCright, A. M. (2011). Organized climate change denial. In J. S. Dryzek, R. B. Norgaard, & D. Schlosberg (Eds.), *The Oxford handbook of climate change and society* (pp. 144–160). Oxford University Press.

Dunlap, R. E., & Van Liere, K. D. (1978). The "new environmental paradigm." *The Journal of Environmental Education, 9*(4), 10–19.

Eckstein, D., Künzel, V., & Schäfer, L. (2021). *Global climate risk index 2021.* Germanwatch.

Ecocidelaw. (2021). Ecocide law. https://ecocidelaw.com/

Eisenfeld, J., Dominguez, R., & Breit, R. (2022). *Tricks of the trade: Deceptive practices, climate delay, and greenwashing in the oil and gas industry.* Earthworks.

Eisinger, J. (2017). *The chickenshit club: Why the Justice Department fails to prosecute executives.* Simon & Schuster.

Environmental Protection Agency. (2021). *Climate change indicators in the United States.* https://www.epa.gov/climate-indicators

Etzioni, A. (2020, November 9). *Is China a new colonial power: How well do the claims of neocolonialism stand up?* The Diplomat. https://thediplomat.com/2020/11/is-china-a-new-colonial-power/

Fact Sheet. (2022, May 2). Department of Homeland Security. Retrieved May 6, 2022, from https://www.dhs.gov/news/2022/05/02/fact-sheet-dhs-internal-working-group-protects-free-speech-other-fundamental-rights

Fairbrother, M. (2016). Externalities: Why environmental sociology should bring them in. *Environmental Sociology, 2*(4), 375–384. http://dx.doi.org/10.1080/23251042.2016.1196636.

Farrell, J. (2016). Corporate funding and ideological polarization about climate change. *Proceedings of the National Academy of Sciences, 113*(1), 92–97.

Farrell, J., McConnell, K., & Brulle, R. (2019). Evidence-based strategies to combat scientific misinformation. *Nature Climate Change, 9*(3), 191–195.

Federici, S. (2018). *Re-enchanting the world: Feminism and the politics of the commons.* PM Press.

Ferrante, L., & Fearnside, P. M. (2019). Brazil's new president and 'ruralists' threaten Amazonia's environment, traditional peoples and the global climate. *Environmental Conservation, 46*(4), 261–263. https://doi.org/10.1017/S0376892919000213.

Fester, J., & Valenzuela, J. (2021). *Environmental science for grades 6–12: A project-based approach to solving the earth's most urgent problems.* International Society for Technology in Education.

Finkel, E. J., Bail, C. A., Cikara, M., Ditto, P. H., Iyengar, S., Klar, S., Mason, L., McGrath, M. C., Nyhan, B., Rand, D. G., Skitka, L. J., Tucker, J. A., Van Bavel, J. J., Wang, C. S., & Druckman, J. N. (2020). Political sectarianism in America. *Science, 370*(6516), 533–536. https://doi.org/ 10.1126/science.abe1715

Fischer, F. (2019). Knowledge politics and post-truth in climate denial: On the social construction of alternative facts. *Critical Policy Studies, 13*(2), 133–152. doi:10.1080/19460171.2019.1602067

Fish, S. (2001). Holocaust denial and academic freedom. *Valparaiso University Law Review, 35*(3), 499–524.

Flanagin, A. J., & Metzger, M. J. (2010). *Kids and credibility: An empirical examination of youth, digital media use, and information credibility.* The MIT Press.

Flavelle, C. (2021, September 2021). Climate change is bankrupting America's small towns. *The New York Times.* https://www.nytimes.com/2021/09/02/climate/climate-towns-bankruptcy.html

Franta, B. (2021). Weaponizing economics: Big Oil, economic consultants, and climate policy delay. *Environmental Politics.* https://doi.org/10.1080/09644016.2021.1947636

Fraser, N. (2017, November 20). From progressive neoliberalism to Trump—and beyond. *American Affairs, 1*(4), 46–64. https://americanaffairsjournal.org/2017/11/progressive-neoliberalism-trump-beyond/

Freese, B. (2020). *Industrial-strength denial: Eight stories of corporations defending the indefensible, from the slave trade to climate change.* University of California Press.

Freud, A. (1936). *The ego and the defense mechanisms.* The Institute of Psychoanalysis.

Friedman, L., & Davenport, C. (2021, August 13). Amid extreme weather, a shift among Republicans on climate change. *The New York Times.* https://www.nytimes.com/2021/08/13/climate/republicans-climate-change.html

Friedrich, J., Ge, M., & Pickens, A. (2020, December 10). *This interactive chart explains the world's top 10 emitters.* World Resources Institute. https://www.wri.org/insights/interactive-chart-shows-changes-worlds-top-10-emitters

Friends of the Earth. (2021, September 16). *Four days of Texas-sized disinformation: Social media companies threaten action on climate change.* https://foe.org/resources/four-days-of-texas-sized-disinformation/

Frosch, R. M., Pastor, M., Sadd, J., & Shonkoff, S. (2018). The climate gap: Inequalities in how climate change hurts Americans and how to close the gap. In E. M. Hamin Infield, Y. Anunnasr, & R. L. Ryan (Eds.), *Planning for climate change: A reader in green infrastructure and sustainable design for resilient cities* (pp. 138–150). Routledge.

Gardiner, B. (2020, June 9). *Unequal impact: The deep links between racism and climate change.* Yale Environment 360. https://e360.yale.edu/features/unequal-impact-the-deep-links-between-inequality-and-climate-change

Garrett, H. J. (2017). *Learning to be in the world with others: Difficult knowledge and social studies education.* Peter Lang.

Garrett, H. J. (2019). Learning to tolerate the devastating realities of climate crises. *Theory & Research in Social Education, 47*(4), 609–614. doi: 10.1080/00933104.2019.1656989

Geertz, C. (1975). On the nature of anthropological understanding: Not extraordinary empathy but readily observable symbolic forms enable the anthropologist to grasp the unarticulated concepts that inform the lives and cultures of other peoples. *American Scientist, 63*(1), 47–53.

Gesturing Towards Decolonial Futures. (n.d). https://decolonialfutures.net/4denials/

Giachino, A., & Mehta-Neugebauer, R. (2021). *Private equity propels the climate crisis: The risks of a shadowy industry's massive exposure to oil, gas and coal.* Private Equity Stakeholder Project. https://www.ourenergypolicy.org/resources/private-equity-propels-the-climate-crisis-the-risks-of-a-shadowy-industrys-massive-exposure-to-oil-gas-and-coal/

Gibson, M. (2020). From deliberation to counter-narration: Toward a critical pedagogy for democratic citizenship. *Theory & Research in Social Education, 48*(3), 431–454. https://doi.org/10.1080/00933104.2020.1747034

Gilbert, D. T. (1991). How mental systems believe. *American Psychologist, 46*(2), 107–119.

Goldberg, M. H., Marlon, J. R., Wang, X., van der Linden, S., & Leiserowitz, A. (2020). Oil and gas companies invest in legislators that vote against the environment. *Proceedings of the National Academy of Sciences, 11710,* 5111–5112. https://doi.org/10.1073/pnas.1922175117

Goodell, J. (2021, June 9). What to do about Jair Bolsonaro, the world's most dangerous climate denier. *Rolling Stone.* https://www.rollingstone.com/politics/politics-features/jair-bolsonaro-rainforest-destruction-1180129/

Goonatilake, S. (1998). *Toward global science: Mining civilizational knowledge.* Indiana University Press.

Gordenker, L. (2017). *The United Nations in international politics.* Princeton University Press

Graeber, D. (2008). *Hope in common.* The Anarchist Library. https://theanarchistlibrary.org/library/david-graeber-hope-in-common

Graeber, D., & Wengrow, D. (2021). *The dawn of everything: A new history of humanity.* Allen Lane.

Graff, M., & Carley, S. (2020, May 1). COVID-19 assistance needs to target energy insecurity. *Nature Energy, 5*, 352–354. https://www.nature.com/articles/s41560 -020-0620-y

Griffin, P. (2017). *The Carbon Majors Database: CDP Carbon Majors Report 2017.* CDP. https://cdn.cdp.net/cdp-production/cms/reports/documents/000/002/327/original /Carbon-Majors-Report-2017.pdf?1501833772

Gross, S. (2020, August 4). *What is the Trump administration's track record on the environment?* Policy 2020 Brookings VoterVitals. https://www.brookings.edu /policy2020/votervital/what-is-the-trump-administrations-track-record-on-the -environment/

Guess, A., Nyhan, B., & Reifler, J. (2020). Exposure to untrustworthy websites in the 2016 U.S. election. *Nature Human Behaviour, 4*(5), 472–480. https://doi.org /10.1038/s41562-020-0833-x

Guterres, A. (2020). *World Social Report 2020: Inequality in a rapidly changing world. United Nations.* Foreword. https://www.un.org/development/desa/dspd /wp-content/uploads/sites/22/2020/02/World-Social-Report2020-FullReport.pdf

Haidt, J. (2012). *The righteous mind: Why good people are divided by politics and religion.* Pantheon/Random House.

Han, B.-C. (2015). *The burnout society.* Stanford University Press.

Hancock, K. J., & Vivoda, V. (2014). International political economy: A field born of the OPEC crisis returns to its energy roots. *Energy Research & Social Science,* 1(March 2014), 206–216.

Harvey, D. (2004). The 'new' imperialism: Accumulation by dispossession. *Socialist Register, 40,* 63–87.

Harvey, D. (2006). Neo-liberalism as creative destruction. *The Annals of the American Academy of Political and Social Science, 610,* 22–44.

Hauerwas, L. B., Kerkhoff, S. N., & Schneider, S. B. (2021). Glocality, reflexivity, interculturality, and worldmaking: A framework for critical global teaching. *Journal of Research in Childhood Education, 35*(2), 185–199. doi: 10.1080/02568543 .2021.1900714

Hawken, P. (Ed.). (2017). *Drawdown: The most comprehensive plan ever proposed to reverse global warming.* Penguin.

Hayes, K. (2016, October 29). Remember this when you talk about Standing Rock. Yes! https://www.yesmagazine.org/orphan/2016/10/29/how-to-talk-about-standing-rock

Hayhoe, K. (2021). *Saving us: A climate scientist's case for hope and healing in a divided world.* One Signal.

Hemmer, N. (2016). *Messengers of the right: Conservative media and the transformation of American politics.* University of Pennsylvania Press.

Henderson, J., & Drewes, A. (Eds.). (2020). *Teaching climate change in the United States.* Routledge.

Henderson, H., Lickerman, J., & Flynn, P. (2000). *Calvert-Henderson quality of life indicators,* Calvert Group. ISBN 978-0-9676891-0-4

Hess, D. (2009). *Controversy in the classroom: The democratic power of discussion.* Routledge.

Hess, D. & McAvoy, P. (2017). *The political classroom: Evidence and ethics in democratic education.* Routledge.

Hester, R. (2010). *Design for ecological democracy.* Retrieved from http://www .radicaldemocracy.org/

Hestres, L. E. (2020). Fighting climate change denial in the United States. In N. Almiron & J. Xifra (Eds.), *Climate change denial and public relations: Strategic communication and interest groups in climate inaction* (pp. 217–232). Routledge.

Hirschfeld, A. (2021, October 2). *Joe Manchin has made $5.2M from his coal company—and gets big donations from fossil-fuel industry*. Salon. https://www.salon .com/2021/10/02/joe-manchin-has-made-52m-from-his-coal-company--and-gets -big-donations-from-fossil-fuel-industry_partner/

Hofstadter, R. (1964). The paranoid style in American politics. *Harper's Magazine*, November 1964, 77–86.

Hoofnagle, M. (2007, April 30). Hello *Scienceblogs*. *Scienceblogs*. https://scienceblogs .com/denialism/2007/04/30/hello-to-scienceblogs

Houser, N. (2009) Ecological democracy: An environmental approach to citizenship education. *Theory & Research in Social Education, 37*(2), 192–214. https://doi .org/ 10.1080/00933104.2009.10473394

Idso, C. D., Carter, R. M., & Singer, S. F. (2015). *Why scientists disagree about global warming: The NIPCC report on scientific consensus*. The Heartland Institute.

IEA. (2021). Net zero by 2050: A roadmap for the global energy sector. https://www .iea.org/reports/net-zero-by-2050

Immordino-Yang, M., & Damasio, A. (2007). We feel, therefore we learn: The relevance of affective and social neuroscience to education. *Mind, Brain, and Education 1*(1), 3–10.

Inglehart, R. F. (1977). *The silent revolution: Changing values and political styles among Western publics*. Princeton University Press.

InfluenceMap. (2019, March). *Big oil's real agenda on climate change*. https://influencemap .org/report/how-big-oil-continues-to-oppose-the-paris-agreement-38212275958aa 21196dae3b76220bddc

InfluenceMap. (2021, August). *Climate change and digital advertising: The oil and gas industry's digital advertising strategy*. https://influencemap.org/report/Climate .-Change-and-Digital-Advertising-a40c8116160668aa2d865da2f5abe91b#

IPCC. (2018). *Global warming of 1.5°C* [Special report]. IPCC. Available from: www .ipcc.ch/sr15.

IPCC. (2020). Climate change and land: An IPCC special report on climate change, desertification, land degradation, sustainable land management, food security, and greenhouse gas fluxes in terrestrial ecosystems. https://www.ipcc.ch/srccl/

IPCC. (2021). Sixth assessment report. https://www.ipcc.ch/assessment-report/ar6/

Jackson, T. (2021). *Postgrowth: Life after capitalism*. John Wiley & Sons.

Jahren, H. (2020). *The story of more: How we got to climate change and where to go from here*. Vintage Books.

Jamieson, K. H. (2008). *Echo chamber: Rush Limbaugh and the conservative media establishment*. Oxford University Press.

Jefferson, T. (1809). Thomas Jefferson to the Republicans of Washington County, Maryland, 31 March 1809. *Founders Online,* National Archives. https://founders .archives.gov/documents/Jefferson/03-01-02-0088.

Jensen, T. (2019). *Ecologies of guilt in environmental rhetorics*. Palgrave Macmillan.

Jesdale, B. M., Morello-Frosch, R., & Cushing, L. (2013). The racial/ethnic distribution of heat risk–related land cover in relation to residential segregation. *Environmental Health Perspectives, 121*(7), 811–817.

Johnson, A. E., & Wilkinson, K. K. (Eds.). (2020). *All we can save: Truth, courage, and solutions for the climate crisis.* One World. https://www.allwecansave.earth/story

Johnson, J. (2021). *Analysis shows Facebook allows 99% of climate disinformation to go unchecked.* Common Dreams. https://www.commondreams.org/news/2021/09/16/analysis-shows-facebook-allows-99-climate-disinformation-go-unchecked

Journell, W. (2017). Fake news, alternative facts, and Trump: Teaching social studies in a post-truth era. *Social Studies Journal, 37,* 8–21.

Kagubare, I. (2019, February 8). Some states still lag in teaching climate science. *Scientific American.* https://www.scientificamerican.com/article/some-states-still-lag-in-teaching-climate-science/

Kahan, D. M. (2017). Misinformation and identity-protective cognition (October 2, 2017). Yale Law & Economics Research Paper No. 587. *SSRN.* http://dx.doi.org/10.2139/ssrn.3046603

Kahn, R. V. (2010). *Critical pedagogy, ecoliteracy, and planetary crisis: The ecopedagogy movement.* Peter Lang.

Kahne, J., & Bowyer, B. (2017). Educating for democracy in a partisan age. *American Educational Research Journal, 54*(1), 3–34. http://dx.doi.org/10.3102/0002831216679817

Kahneman, D. (2011). *Thinking, fast and slow.* Farrar, Straus and Giroux.

Kahn-Harris, K. (2018). *Denial: The unspeakable truth.* Notting Hill Editions.

Kalhoefer, K. (2016, April 26). *CNN viewers see far more fossil fuel advertising than climate reporting.* Media Matters for America—EcoWatch. https://www.ecowatch.com/cnn-viewers-see-far-more-fossil-fuel-advertising-than-climate-reportin-1891124850.html

Kaminski, I. (2018, November 19). *Philippines' typhoon survivors to fossil fuel firms: Please listen to us.* Eco-business. https://www.eco-business.com/news/philippines-typhoon-survivors-to-fossil-fuel-firms-please-listen-to-us/

Kavanagh, J., & Rich, M. D. (2018). *Truth decay: An initial exploration of the diminishing role of facts and analysis in American public life.* Rand Corporation.

Keep it in the ground. (n.d.). The sky's limit: No new fossil fuel development: An open letter to world leaders. #KeepItInTheGround. http://keepitintheground.org/#read-the-letter

Kelly, S. (2021, August 5). *Oil and gas inundated Facebook with election season ads after Biden released climate plan.* DeSmog. https://www.desmog.com/2021/08/05/oil-gas-facebook-election-ads-biden-climate/

Kimmerer, R. W. (2013). *Braiding sweetgrass: Indigenous wisdom, scientific knowledge and the teachings of plants.* Milkweed Editions.

Kirsch, A., Opena Disterhoft, J., Marr, G., McCully, P., Breech, R., Dilworth, T., . . . & Wickham, M. S. (2021). Banking on Climate Chaos 2021. (INIS-FR-21-0705). France. https://www.ran.org/publications/banking-on-climate-chaos-2021/

Kissling, M. T., & Bell, J. T. (2019). Teaching social studies amid ecological crisis. *Theory & Research in Social Education, 48*(1), 1–31. https://doi.org/10.1080/00933104.2019.1673267

Kissling, M. T., Bell, J. T., Díaz Beltrán, A. C., & Myler, J. L. (2017). Ending the silence about the Earth in social studies teacher education. In C. C. Martell (Ed.), *Social studies teacher education: Critical issues and current perspectives* (pp. 193–220). Information Age.

Klein, N. (2011, November 28). Capitalism vs. the climate. *The Nation.* https://www
.thenation.com/article/archive/capitalism-vs-climate/

Klein, N. (2014). *This changes everything: Capitalism vs the climate.* Simon & Schuster.

Klein, N., & Stefoff, R. (2021). *How to change everything: The young human's guide
to protecting the planet and each other.* Penguin.

Korten, D. C. (2007). *The great turning: From empire to earth community.* Berrett-Koehler.

Kraft, P. W., Lodge, M., & Taber, C. S. (2014). Why people 'don't trust the evidence':
Motivated reasoning and scientific beliefs. *The ANNALS of the American Academy of Political and Social Science, 658,* 121–133.

Kroll, D. (2013, Feb 5). Exposed: The dark-money ATM of the conservative movement. *Mother Jones.* https://www.motherjones.com/politics/2013/02/donors
-trust-donor-capital-fund-dark-money-koch-bradley-devos/

Krutka, D. (Host). (2020, February 8). Teaching social studies amid ecological crisis with Mark Kissling, Jonathan Bell with guests Jim Garrett & Zack Seitz.
(No. 135) [Audio podcast episode]. *Visions of Education.* https://visionsofed.com
/2020/02/08/episode-135-teaching-social-studies-amid-ecological-crisis-with
-mark-kissling-jonathan-bell-with-guests-jim-garrett-zack-seitz/

Kull, S., Ramsay, C., & Lewis, E. (2003). Misperceptions, the media, and the Iraq
war. *Political Science Quarterly, 118*(4), 569–598.

Kusnetz, N. (2020). What does net zero emissions mean for big oil? Not what you'd
think. *Inside Climate News.* https://insideclimatenews.org/news/16072020/oil
-gas-climate-pledges-bp-shell-exxon/

Lakoff, G. (2014). *Don't think of an elephant!: Know your values and frame the debate: the essential guide for progressives.* Chelsea Green Publishing.

Lakoff, G. (2017, January 12). *A taxonomy of Trump tweets.* On the Media. WYNC Stu-
dios. https://www.wnycstudios.org/podcasts/otm/segments/taxonomy-trump-tweets

Lamb, W. F., Mattioli, G., Levi, S., Roberts, J. T., Minx, J. C., Mueller-Hansen, F.,
Capstick, S., Creutzig, F., Culhane, T., and Steinberger, J. K. (2020). Discourses of
climate delay. *Global Sustainability, 3,* 1–5. https://doi.org/10.1017/sus.2020.13

Lankshear, C., & Knobel, M. (2003). *New literacies: Changing knowledge and classroom learning.* Open University Press.

Larsen, K., Pitt, H., Grant, M., & Houser, T. (2021, May 6). *China's greenhouse gas
emissions exceeded the developed world for the first time in 2019.* Rhodium
Group. https://rhg.com/research/chinas-emissions-surpass-developed-countries/

Lasch, C. (1979). *The culture of narcissism: American life in an age of diminishing
expectations.* W. W. Norton.

Lasch, C. (1991). *The true and only heaven: Progress and its critics.* W. W. Norton
& Company.

LastWeekTonight (2014, May 11). *Climate change debate: Last week tonight with
John Oliver* (HBO) [Video]. YouTube. https://www.youtube.com/watch?v
=cjuGCJJUGsg

Latour, B. (2018). *Down to earth: Politics in the new climatic regime* (C. Porter,
Trans.). Polity.

Leiserowitz, A., Maibach, E. W., & Roser-Renouf, C. (2009). Climate change in the
American mind: Americans' climate change beliefs, attitudes, policy preferences,
and actions. *SSRN.* http://dx.doi.org/10.2139/ssrn.2667029

Leiserowitz, A., Marlon, J., Wang, X., Bergquist, P., Goldberg, M., Kotcher, J., Maibach, E., & Rosenthal, S. (2020). *Global warming's six Americas in 2020.* Yale

Program on Climate Change Communication. https://climatecommunication
.yale.edu/publications/global-warmings-six-americas-in-2020/

Lepore, J. (2021). *The last archive* [Podcast]. https://www.thelastarchive.com/

Levy, B. S., & Patz, J. A. (2015). Climate change, human rights, and social justice. *Annals of Global Health, 81*(3), 310–322. https://doi.org/10.1016/j.aogh
.2015.08.008

Lewandowsky, S., Cook, J., Oberauer, K., Brophy, S., Lloyd, E. A., & Marriott, M.
(2015). Recurrent fury: Conspiratorial discourse in the blogosphere triggered by
research on the role of conspiracist ideation in climate denial. *Journal of Social
and Political Psychology, 3*(1), 142–178.

Lewandowsky, S., Ecker, U., & Cook, J. (2017). Beyond misinformation: Understanding and coping with the post-truth era. *Journal of Applied Research in Memory
and Cognition, 6*(4), 353–369.

Lewis, A. (Director & Producer), Barnes, J., & Cuarón, A. (Producers). (2015). *This
changes everything* [Motion picture]. Klein Lewis Productions & Louverture Films.

Litonju, M. D. (2012). Third world/global south: From modernization, to dependency
/liberation, to postdevelopment. *Journal of Third World Studies, 29*(1), 25–56.

Liu, Z. (2005). Reading behavior in the digital environment: Changes in reading behavior over the past ten years. *Journal of Documentation, 61*(6), 700–712.

Longino, H. E. (2020). *Science as social knowledge*. Princeton University Press.

Lopez, A. (2021). *Ecomedia literacy: Integrating ecology into media education*. Routledge.

Löwy, M. (2015). *Ecosocialism: A radical alternative to capitalist catastrophe*. Hay-
market Books.

Luke, A. (2014). Defining critical literacy. In J. Z. Pandya & J. Àvila (Eds.), *Moving critical literacies forward: A new look at praxis across contexts* (pp. 19–31).
Routledge.

Macy, J., & Johnstone, C. (2012). *Active hope: How to face the mess we're in without
going crazy*. New World Library.

Manji, I. (2019). *Don't label me: An incredible conversation for divided times*.
St. Martin's Press.

Mann, M. (2012). *The hockey stick and the climate wars*. Columbia University Press.

Mann, M. (2021). *The new climate war: The fight to take back the planet*. Public Affairs.

Marchi, R. (2012). With Facebook, blogs, and fake news, teens reject journalistic
"objectivity." *Journal of Communication Inquiry, 36*(3), 246–262.

Marks, E., Hickman, C., Pihkala, P., Clayton, S., Lewandowski, E. R., Mayall, E. E.,
Wray, B., Mellor, C., & van Susteren, L. (2021, September 7). Young people's
voices on climate anxiety, government betrayal and moral injury: A global phe-
nomenon. *SSRN*. http://dx.doi.org/10.2139/ssrn.3918955

Marshall, G. (2015). *Don't even think about it: Why our brains are wired to ignore
climate change*. Bloomsbury.

Martens, H., & Hobbs, R. (2015). How media literacy supports civic engagement in
a digital age. *Atlantic Journal of Communication, 23*(2), 120–137.

Martin, J. (2007). *The meaning of the 21st century: A vital blueprint for ensuring our
future*. Riverhead Penguin.

Martin, J. R. (1994). *Changing the educational landscape: Philosophy, women, and
curriculum*. Routledge.

Martusewicz, R. A., Edmundson, J., & Lupinacci, J. (2021). *Ecojustice education:
Toward diverse, democratic, and sustainable communities* (3rd ed.). Routledge.

Mason, P. (2015). Postcapitalism: A guide to our future. Allen Lane.

Massey, D. (1994). *Space, place, and gender*. University of Minnesota Press.

Masyada, S., & Washington, E. Y. (2016). Civil Liberties, media literacy, and civic education in the post-9/11 era: Helping students think conceptually in order to act civically. In W. Journell (Ed.), *Reassessing the social studies curriculum: Promoting critical civic engagement in a politically polarized, post-9/11 world* (pp. 83–94). Rowman & Littlefield.

Maturana, H. R. (2008). *The origin of humanness in the biology of love*. Imprint Academic.

Maximillian, J., Brusseau, M. L., Glenn, E. P., & Matthias, A. D. (2019). Chapter 25: pollution and environmental perturbations in the global system. In M. L. Brusseau, I. L. Pepper, & C. P. Gerba (Eds.), *Environmental and Pollution Science* (3rd ed.; pp. 457–476). Academic Press. https://doi.org/https://doi.org/10.1016/B978-0-12-814719-1.00025-2

Max-Neef, M. (1992). *From the outside looking in: Experiences in barefoot economics*. Zed Books.

Mayer, J. (2016). *Dark money: The hidden history of the billionaires behind the rise of the radical right*. Anchor Books.

Mayo, R., & Novack, R. (n.d.). *ELATE Commission on climate change and the environment in English education (c3e3)*. English Language Arts Teacher Educators, National Council of Teachers of English. Retrieved from: https://ncte.org/groups/elate/elate-commissions/

McGhee, H. (2021). *The sum of us: What racism costs everyone and how we can prosper together*. One World.

McKay, A. (Director). (2021). *Don't look up* [Film]. Hyperobject Industries/Bluegrass Films.

McKeown, M. G., & Beck, I. L. (1994). Making sense of accounts of history: Why young students don't and how they might. In O. Hallden, G. Leinhardt, I.L. Beck, & C. Stainton (Eds.), *Teaching and learning in history* (pp. 1–26). Lawrence Erlbaum.

McKewon, E. (2012). Talking points ammo: The use of neoliberal think tank fantasy themes to delegitimise scientific knowledge of climate change in Australian newspapers. *Journalism Studies, 13*, 277–297. http://doi.org/10.1080/1461670X.2011.646403

McKibben, B. (2021, August 25). Slow-walking the climate crisis: "Greenwashing" is too kind a term; this is more like careful sabotage. *The New Yorker*. https://www.newyorker.com/news/annals-of-a-warming-planet/slow-walking-the-climate-crisis

Meadows, D. H., & Randers, J. (2013). *Limits to growth*. Chelsea Green.

Michaels, L., & Ainger, K. (2020). The climate smokescreen: The public relations consultancies working to obstruct greenhouse gas emissions reductions in Europe—A critical approach 1. In N. Almiron, & J. Xifra (Eds.), *Climate change denial and public relations* (pp. 159–177). Routledge.

Middaugh, E. (2017, November). *When media literacy meets issue advocacy: Adolescents' use of media literacy strategies in civic inquiry* [Paper presentation]. Annual meeting of the College and University Faculty Assembly of the National Council for the Social Studies, San Francisco, CA.

Mills, C. W. (1956). *The power elite*. Oxford University Press.

Milman, O. (2020, February 21). Revealed: Quarter of all tweets about climate crisis produced by bots. *The Guardian*. https://www.theguardian.com/technology/2020/feb/21/climate-tweets-twitter-bots-analysis

Monbiot, G. (2017). *Out of the wreckage: A new politics in an age of crisis*. Verso.

Monbiot, G. (2021, November 10). Make extreme wealth extinct: It's the only way to avoid climate breakdown. *The Guardian*. https://www.theguardian.com/commentisfree/2021/nov/10/extreme-wealth-polluting-climate-breakdown-rich

Mooney, H. A., Duraiappah, A., & Larigauderie, A. (2013). Evolution of natural and social science interactions in global change research programs. In J. Shaman (Ed.), *Proceedings of the National Academy of Sciences, Columbia University, New York, NY, 110* (Supplement 1), 3665–3672.

Mouffe, C. (2013). *Agonistics: Thinking the world politically*. Verso Books.

Mouffe, C. (2018). Agonistic democracy and radical politics. *Pavilion*. http://pavilionmagazine.org/chantal-mouffe-agonistic-democracy-and-radical-politics/

Muller, M. (2020). In search of the global east: Thinking between north and south. *Geopolitics, 25*(3), 734–755.

Myers, K. F., Doran, P. T., Cook, J., Kotcher, J. E., & Myers, T. A. (2021). Consensus revisited: quantifying scientific agreement on climate change and climate expertise among Earth scientists 10 years later. *Environmental Research Letters, 16*(10), 104030.

Nader, R. (2016a). *Animal envy*. Seven Stories Press.

Nader, R. (2016b). *Breaking through power: It's easier than we think*. City Lights.

Nader, R. (2019, April 24). Children's moral power can challenge corporate power on climate crisis. *The Litchfield County Times*. http://www.countytimes.com/opinion/ralph-nader-children-s-moral-power-can-challenge-corporate-power-on-climate-crisis/article_a5eb5f80-58ec-5874-85de-529f01364351.html

Naess, A. (1995). The shallow and the long range, deep ecology movement. In A. Drengson and Y. Inoue (Eds.), *The deep ecology movement: An introductory anthology* (pp. 3–10). North Atlantic Books.

Nagel, J. (2016). *Gender and climate change: Impacts, science, policy*. Routledge.

Nakate, V. (2022, April 8). This 900-mile crude oil pipeline is a bad deal for my country—and the world. *The New York Times*. https://www.nytimes.com/2022/04/08/opinion/environment/east-africa-oil-pipeline.html?referringSource=articleShare

NAMLE (National Association for Media Literacy). (2007, November). Core principles of media literacy education in the United States. Retrieved January 7, 2017, from https://namle.net/publications/core-principles.

National Council for the Social Studies (NCSS). (2013). *College, career and civic life (C3) framework for social studies state standards: Guidance for enhancing the rigor of K-12 civics, economics, geography and history*. National Council for the Social Studies.

National Council for the Social Studies (NCSS). (2019). *Teaching climate change: Updating online resources* [Resolution #19-02-2]. https://www.socialstudies.org/about/hod/2019-hod-resolutions#19-02-2

National Council of Teachers of English (NCTE). (2019). *Resolution on literacy teaching on climate change* [Position statement]. https://ncte.org/statement/resolution-literacy-teaching-climate-change/

National Integrated Climate Change Database and Information Exchange System (NICCDIES). (2021). *Climate change impacts*. https://niccdies.climate.gov.ph/climate-change-impacts

Nelson, M. (2021). *On freedom: Four songs of care and constraint.* Graywolf Press.

Newberry, T., & Trujillo, O. V. (2019). Decolonizing education through transdisciplinary approaches to climate change education. In L. T. Smith, E. Tuck, & K. W. Yang (Eds.), *Indigenous and decolonizing studies in education: Mapping the long view* (pp. 204–214). Routledge.

Nicholas, K. (2021). *Under the sky we make: How to be human in a warming world.* G. P. Putnam's Sons.

Nichols, J., & McChesney, R. W. (2010, November 10). The money and media election complex. *The Nation.* https://www.thenation.com/article/archive/money-media-election-complex/

Nichols, T. (2017). *The death of expertise: The campaign against established knowledge and why it matters.* Oxford University Press.

Nicolaidou, I., Kyza, E. A., Terzian, F., Hadjichambis, A., & Kafouris, D. (2011). A framework for scaffolding students' assessment of the credibility of evidence. *Journal of Research in Science Teaching, 48*(7), 711–744.

Norgaard, K. M. (2011). *Living in denial: Climate change, emotions, and everyday life.* The MIT Press.

Nussbaum, M. C. (2011). *Creating capabilities: The human development approach.* Harvard University Press.

Nyberg, D. (2021). Corporations, politics, and democracy: Corporate political activities as political corruption. *Organization Theory, 2,* 1–24. https://journals.sagepub.com/doi/10.1177/2631787720982618

Nyhan, B. (2014, May 20). The role of elites in Holocaust denial. *The New York Times.* https://www.nytimes.com/2014/05/21/upshot/the-role-of-elites-in-holocaust-denial.html

Nyhan, B., & Reifler, J. (2010). When corrections fail: The persistence of political misperceptions. *Political Behavior, 32*(2), 303–330.

O'Brien, K. J. (2017). *The violence of climate change: Lessons of resistance from nonviolent activists.* Georgetown University Press.

Ojala, M. (2017). Hope and anticipation in education for a sustainable future. *Futures, 94,* 76–84. https://www.sciencedirect.com/science/article/pii/S0016328716301422?via%3Dihub

O'Neill, L. (2019, May 29). US energy department rebrands fossil fuels as 'molecules of freedom'. *The Guardian.* https://www.theguardian.com/business/2019/may/29/energy-department-molecules-freedom- fossil-fuel-rebranding

Oreskes, N. (2019). *Why trust science?* Princeton University Press.

Oreskes, N., & Conway, E. M. (2010). *Merchants of doubt: How a handful of scientists obscured the truth on issues from tobacco smoke to global warming.* Bloomsbury.

Orr, D. W. (1992). *Ecological literacy: Education and the transition to a postmodern world.* State University of New York Press.

Orr, D. W. (1994). *Earth in mind: On education, environment, and the human prospect.* Island Press.

Osborne, M. (2022, April 13). Scientists stage worldwide climate change protests after IPCC report. *Smithsonian Magazine.* https://www.smithsonianmag.com/smart-news/scientists-stage-worldwide-climate-protests-after-ipcc-report-180979913/

Oxfam International. (2015, December 2). *Extreme carbon inequality: Why the Paris climate deal must put the poorest, lowest emitting and most vulnerable people first* [Oxfam media briefing].

Oxford Martin School. (2021). *Our world in data: Per capita meat consumption by type, 2017.* https://ourworldindata.org/grapher/per-capita-meat-type?country=CHN~USA~IND~ARG~PRT~ETH~JPN~GBR~BRA

Pace, J. L. (2019). Contained risk-taking: Preparing preservice teachers to teach controversial issues in three countries. *Theory & Research in Social Education, 47*(2), 228–260. doi:10.1080/00933104.2019.1595240

Painter, J. (2011). *Poles apart: The international reporting of climate scepticism.* Reuters Institute for the Study of Journalism, University of Oxford.

Panchyk, R. (2021). *Power to the people! A young people's guide to fighting for our rights as citizens and consumers.* Seven Stories Press.

Panos, A., & Damico, J. S. (2021). Less than one percent is not enough. How leading literacy organizations engaged with climate change from 2008 to 2019. *Journal of Language and Literacy Education, 17*(1), 1–21.

Parker, W. (2011). Foreword. In M. Baildon & J. S. Damico, *Social studies as new literacies in a global society: Relational cosmopolitanism in the classroom* (pp. xiii–xv). Routledge/Taylor & Francis.

Parker, W. C. (2006). Public discourses in schools: Purposes, problems, possibilities. *Educational Researcher, 35*(8), 11–18.

Parry, I., Black, S., & Vernon, N. (2021). *Still not getting energy prices right: A global and country update of fossil fuel subsidies.* International Monetary Fund Working Papers. https://www.imf.org/en/Publications/WP/Issues/2021/09/23/Still-Not -Getting-Energy-Prices-Right-A-Global-and-Country-Update-of-Fossil-Fuel -Subsidies-466004

Paul, K. (2021, November 4). Climate misinformation on Facebook "increasingly substantially," study says. *The Guardian.* https://www.theguardian.com/technology /2021/nov/04/climate-misinformation-on-facebook-increasing-substantially-study -says

Persily, N. (2019). *Elections and democracy in the digital age.* Kofi Annan Foundation.

Peters, M. A. (2017). Education for ecological democracy. *Educational Philosophy and Theory, 49*(10), 941–945. doi:10.1080/00131857.2017.1339408

Peterson-Withorn, C. (2021, April 30). How much money America's billionaires have made during the Covid-19 pandemic. *Forbes.* https://www.forbes.com/sites /chasewithorn/2021/04/30/american-billionaires-have-gotten-12-trillion-richer -during-the-pandemic/?sh=70ff8590f557

Pilon, M. (2021, February 6). *Big oil gets to teach climate science in American classrooms.* Bloomberg Green. https://www.bloomberg.com/news/features/2021-02 -06/big-oil-gets-to-teach-climate-change-in-american-classrooms

Pollin, R. (2022, April 8). *Nationalize the U.S. fossil fuel industry to save the planet.* The American Prospect. https://prospect.org/environment/nationalize-us-fossil -fuel-industry-to-save-the-planet/

Pope Francis. (2015). *Laudato si' of the Holy Father Francis on care for our common home* [Encyclical]. St Pauls Publications.

Prádanos, L. I. (2018). *Postgrowth imaginaries: New ecologies and counterhegemonic culture in post-2008 Spain.* Liverpool University Press.

Prior, M. (2013). Media and political polarization. *Annual Review of Political Science, 16*(1), 101–127. doi:10.1146/annurev-polisci-100711-135242

Randall, R. (2005). A new climate for psychotherapy? *Psychotherapy and Politics International, 3*(3), 165–179.

Rauch, J. (2021). *The constitution of knowledge: A defense of truth*. Brookings Institution Press.

Raworth, K. (2017). *Doughnut economics: Seven ways to think like a 21st-century economist*. Chelsea Green.

Real solutions not "net zero." (n.d.). A global call for climate action. https://www.realsolutions-not-netzero.org/

Regan, H., & Dotto, C. (2021, November 3). *US vs China: How the world's two biggest emitters stack up on climate*. CNN. https://edition.cnn.com/2021/10/28/world/china-us-climate-cop26-intl-hnk/index.html

Rich, N. (2019). *Losing earth: A recent history*. Farrar, Straus and Giroux.

Ritchie, H., & Roser, M. (2020). CO_2 *and greenhouse gas emissions*. OurWorldIn Data.org. https://ourworldindata.org/co2-and-other-greenhouse-gas-emissions

Roberts, D. (2017, June 14). *Donald Trump is handing the federal government over to fossil fuel interests*. Vox. https://www.vox.com/energy-and-environment/2017/6/13/15681498/trump-government-fossil-fuels

Rogers, A., Castree, N., & Kitchin, R. (2013). *A dictionary of human geography*. Oxford University Press.

Ruitenberg, C. (2010). Conflict, affect and the political: On disagreement as democratic capacity. *In Factis Pax: Journal of Peace Education and Social Justice, 4*(1), 40–55.

Sabzalian, L. (2019). The tensions between Indigenous sovereignty and multicultural citizenship education: Toward an anticolonial approach to civic education. *Theory & Research in Social Education, 47*(3), 311–346. doi:10.1080/00933104.2019.1639572

Sadler, T. (2004). Informal reasoning regarding socioscientific issues: A critical review of research. *Journal of Research in Science Teaching, 41*(5), 513–536.

Saffer POV. (2021, Sept. 23). "Maybe We Didn't Think This Through" oil company ad. [Video]. https://www.youtube.com/watch?v=cAhD0NfuWgk

Sant, E., McDonnell, J., Pashby, K., & Menendez Alvarez-Hevia, D. (2021). Pedagogies of agonistic democracy and citizenship education. *Education, Citizenship and Social Justice, 16*(3), 227–244.

Schlemper, M. B., Stewart, V. C., Shetty, S., & Czajkowski, K. (2018). Including students' geographies in geography education: Spatial narratives, citizen mapping, and social justice. *Theory & Research in Social Education, 46*(4), 603–641. doi:10.1080/00933104.2018.1427164

Schneider, J., Schwarze, S., Bsumek, P. K., & Peeples, J. (2016). *Under pressure: Coal industry rhetoric and neoliberalism*. Palgrave Macmillan.

Schneider, S. H. (1989). *Global warming: Are we entering the greenhouse century?* (Vol. 90). Sierra Club Books.

Schneider-Mayerson, M., & Bellamy, R. (Eds.). (2019). *An ecotopian lexicon*. University of Minnesota Press.

Schumacher, E. F. (1973). *Small Is beautiful: Economics as if people mattered*. Harper & Row.

Segall, A. (2003). Maps as stories about the world. *Social Studies and the Young Learner, 16*(1), 21–25.

Segall, A. (2006). What's the purpose of teaching a discipline anyway? The case of history. In A. Segall, E. E. Heilman, & C. H. Cherryholmes (Eds.), Social studies—the next generation: Re-searching in the postmodern (pp. 125–139). Peter Lang.

Sen, A. (1999). *Development as freedom*. Oxford University Press.

Shapin, S. (1994). *A social history of truth: Civility and science in seventeenth-century England*. University of Chicago Press.

Shepardson, D. P., Roychoudhury, A., & Hirsch, A. S. (Eds.). (2017). *Teaching and learning about climate change: A framework for educators*. Routledge.

Shepherd, K. (2021, February 18). Rick Perry says Texans would accept even longer power outages "to keep the federal government out of their business." *The Washington Post*. https://www.washingtonpost.com/nation/2021/02/17/texas-abbott -wind-turbines-outages/

Shermer, M. (2011, July 1). The believing brain: Why science is the only way out of belief-dependent realism. *Scientific American*. https://www.scientificamerican .com/article/the-believing-brain/

Sherrington, R. (2021, October 7). *Investigation: Majority of directors of world's top insurance companies tied to polluting industries*. DeSmog. https://www .desmog.com/2021/10/07/climate-conflicted-insurance-directors/?utm_source =DeSmog+Weekly+Newsletter

Shiva, V. (2005). *Earth democracy: Justice, sustainability, and peace*. South End Press.

Sinatra, G., & Hofer, B. (2021). *Science denial: Why it happens and what to do about it*. Oxford University Press.

Sinatra, G. M., & Broughton, S. H. (2011). Bridging reading comprehension and conceptual change in science education: The promise of refutation text. *Reading Research Quarterly*, 46(4), 374–393.

Sirota, D. (2021, September 10). The planet can't survive a repeat of Barack Obama's climate denialism. *Jacobin*. https://www.jacobinmag.com/2021/08/climate -denialism-barack-obama-biden-administration-ipcc-report-crisis

Smith, D. (2021). *The state of the world atlas* (10th ed.). Penguin Books.

Sokol, K. C. (2020). Seeking (some) climate justice in state tort law. *Washington Law Review*, 95(3), 1383. https://digitalcommons.law.uw.edu/wlr/vol95/iss3/7

Solnit, R. (2021, August 23). Big oil coined "carbon footprints" to blame us for their greed. Keep them on the hook. *The Guardian*. https://www.theguardian.com /commentisfree/2021/aug/23/big-oil-coined-carbon-footprints-to-blame-us-for -their-greed-keep-them-on-the-hook

Spires, H., Kerkhoff, S. N., & Paul, C. (2021). *Read, write, inquire: Disciplinary literacy in grades 6–12*. Teachers College Press.

Spiro, R. J., Coulson, R. L., Feltovich, P. J., & Anderson, D. (2004). Cognitive flexibility theory: Advanced knowledge acquisition in ill-structured domains. In R. B. Ruddell (Ed.), *Theoretical models and processes of reading* (5th ed.; 602–616). International Reading Association.

Srnicek, N. (2017). *Platform capitalism*. Polity Press.

Stănescu, V. (2020). "Cowgate": Meat eating and climate change denial. In N. Almiron & J. Xifra (Eds.), *Climate change denial and public relations* (pp. 178–194). Routledge. https://doi.org/10.4324/9781351121798

Steffen, A. [@AlexSteffen]. (2017, September 13). *Predatory delay—the knowing choice to worsen a planetary crisis thru delay in order to get even wealthier while they can—is the story*. [Tweet]. https://twitter.com/AlexSteffen/status /908122407595941888

Steffen, W., Richardson, K., Rockström, J., Cornell, S. E., Fetzer, I., Bennett, E. M., Biggs, R., Carpenter, S. R., De Vries, W., De Wit, C., Folke, C., Gerten, D., Heinke, J., Mace, G. M., Persson, L. M., Ramanathan, V., Reyers, B., & Sörlin, S. (2015).

Planetary boundaries: Guiding human development on a changing planet. *Science, 347*(6223). doi: 10.1126/science.1259855

Stein, S. (2021). Reimagining global citizenship education for a volatile, uncertain, complex, and ambiguous (VUCA) world. *Globalisation, Societies and Education, 19*(4), 482–495. https://doi.org/10.1080/14767724.2021.1904212

Stephenson, W. (2015). *What we're fighting for now is each other: Dispatches from the front lines of climate justice.* Beacon Press.

Stibbe, A. (2021). Ecolinguistics: Language, ecology and the stories we live by (2nd ed.). Routledge.

Stiglitz, J. E., Fitoussi, J.-P., & Durand, M. (2019). *Measuring what counts: The global movement for well-being.* The New Press.

Suldovsky, B. (2017). The information deficit model and climate change communication. *Oxford research encyclopedia of climate science.* https://oxfordre.com /climatescience/view/10.1093/acrefore/9780190228620.001.0001/acrefore -9780190228620-e-301

Sunstein, C. (2021). *Liars: Falsehoods and free speech in an age of deception.* Oxford University Press.

Supran, G., & Oreskes, N. (2021). Rhetoric and frame analysis of ExxonMobil's climate change communications. *One Earth, 4*(5), 696–719.

Swan, K., Lee, J., & Grant, S. G. (2018). *Inquiry design model: Building inquiries in social studies.* National Council for the Social Studies (NCSS).

Sze, J. (2020). *Environmental justice in a moment of danger.* University of California Press. https://doi.org/10.1525/9780520971981

Tabuchi, H. (2021, December 16). Google pledged to remove ads from climate denial sites, but many still run. *The New York Times.* https://www.nytimes.com/2021 /12/16/climate/google-climate-denial-ads.html

Taft, M., & Atkin, E. (2021, October 27). *Big oil uses newsletter ads to spread misinformation ahead of big oil misinformation hearing.* Gizmodo. https://gizmodo .com/big-oil-uses-newsletter-ads-to-spread-misinformation-ah-1847946590

Taylor, E., Gillborn, D., & Ladson-Billings, G. (2009). *Foundations of critical race theory in education.* Routledge.

Turner, R. J. (2015). *Teaching for ecojustice: Curriculum and lessons for secondary and college classrooms.* Routledge.

Teo, Y. (2017). Vignettes of poverty versus stories of inequality. *Media Asia,* 1–7. doi :10.1080/01296612.2016.1274707

Thunberg, G. (2021, October 4). "Blah, blah, blah": Youth climate activists slam political inaction at U.N. summit ahead of COP 26. *Democracy Now!* https://www .democracynow.org/2021/10/4/youth_climate_summit_milan_italy

Tong, D. (2020, September 23). *Discussion paper: Big oil reality check: Assessing oil and gas company climate plans.* Oil Change International. http://priceofoil.org /2020/09/23/big-oil-reality-check/

Treen, K.M.D., Williams, H.T.P., & O'Neill, S. J. (2020). Online misinformation about climate change. *WIREs Climate Change, 11*(5) e665. https://doi.org/10.1002/wcc.665

Tuck, E., & Yang, K. W. (2012). Decolonization is not a metaphor. *Decolonization: Indigeneity, Education and Society, 1*, 1–40.

Tufekci, Z. (2019, April 21). Think you're discreet online? Think again [Opinion; The Privacy Project]. *The New York Times.* https://www.nytimes.com/2019/04 /21/opinion/computational-inference.html

Tully, J. (in press). Trust, mistrust and distrust in diverse societies. In D. Karmis & F. Rocher (Eds.), *Trust and distrust in political theory and practice: The case of diverse societies*. McGill-Queens. https://philpapers.org/rec/TULTMA-2

Turner, R. J. (2015). *Teaching for ecojustice: Curriculum and lessons for secondary and college classrooms*. Routledge.

Tutu, D. (2010). Foreword: The fatal complacency. In F. Kagawa & D. Selby (Eds.), *Education and climate change: Living and learning in interesting times* (pp. xv–xvi). Routledge/Taylor & Francis.

van der Linden, S., Panagopoulos, C., Azevedo, F., & Jost, J. T. (2021). The paranoid style in American politics revisited: An ideological asymmetry in conspiratorial thinking. *Political Psychology, 42*(1), 23–51. doi:10.1111/pops.12681

van Dijck, J. (2014). *The culture of connectivity: A critical history of social media*. Oxford University Press.

VanSledright, B. (2010). What does it mean to think historically . . . and how do you teach it? In W. C. Parker (Ed.), *Social studies today: Research and practice.* (pp. 113–120). Routledge.

VanSledright, B. S., & Afflerbach, P. (2005). Assessing the status of historical sources: An exploratory study of eight US elementary students reading documents. In R. Ashby, P. Gordon, & P. Lee (Eds.), *Understanding history: International review of history education 4* (pp. 1–20). Routledge. https://doi.org/10.4324/9780203340929

Vidal, J. (2016, July 27). World's largest carbon producers face landmark human rights case. *The Guardian*. https://www.theguardian.com/environment/2016/jul/27/worlds-largest-carbon-producers-face-landmark-human-rights-case

Warren, K. (1990). The power and the promise of ecological feminism. *Environmental Ethics, 12*(2), 125–146.

Warren, M. E. (2018). Trust and democracy. In U. M. Uslaner (Ed.), *The Oxford handbook of social and political trust* (pp. 75–94). Oxford University Press.

Warzel, C. (2018, February 11). *Infocalypse now: He predicted the 2016 fake news crisis. Now he's worried about an information apocalypse*. Buzzfeed. https://www.buzzfeednews.com/article/charliewarzel/the-terrifying-future-of-fake-news#.cvx3Vqmr2

Washington, H., & Cook, J. (2011). *Climate change denial: Heads in the sand*. Earthscan.

Watts, J., & Doherty, B. (2018, December 9). US and Russia ally with Saudi Arabia to water down climate pledge. *The Guardian*. https://www.theguardian.com/environment/2018/dec/09/us-russia-ally-saudi-arabia-water-down-climate-pledges-un

Weisbrod, K. (2021, August 15). *The repercussions of a changing climate, in 5 devastating charts*. Inside Climate News. https://insideclimatenews.org/news/15082021/climate-change-ipcc-charts/

Weiss, C. (2003). Expressing scientific uncertainty. *Law, probability and risk, 2*, 25–46.

Westervelt, A. (2021, Sept 9). Big oil's 'wokewashing' is the new climate science denialism. *The Guardian*. https://www.theguardian.com/environment/2021/sep/09/big-oil-delay-tactics-new-climate-science-denial

Williams, W. (2021, August 12). How your cup of coffee is clearing the jungle. *The New York Times*. https://www.nytimes.com/2021/08/11/magazine/indonesia-rainforest-coffee.html

Wineburg, S. (1991). Historical problem solving: A study of the cognitive processes used in the evaluation of documentary and pictorial evidence. *Journal of Educational Psychology, 83*, 73–87.

Wineburg, S. (2001). *Historical thinking and other unnatural acts: Charting the future of teaching the past.* Temple University Press.

Wineburg, S., & McGrew, S. (2019). Lateral reading and the nature of expertise: Reading less and learning more when evaluating digital information. *Teachers College Record, 121*(11), 1–40.

Wineburg, S., McGrew, S., Breakstone, J., & Ortega, T. (2016). *Evaluating information: The cornerstone of civic online reasoning.* Stanford Digital Repository. http://purl.stanford.edu/fv751yt5934.

Wolfmeyer, M., Lupinacci, J., & Chesky, N. (2017). EcoJustice mathematics education: An ecocritical (re)consideration for 21st century curricular challenges. *Journal of Curriculum Theorizing, 32*(2), 53–71.

World Health Organization. (2003). WHO framework convention on tobacco control. World Health Organization.

Worth, K. (2021). *Miseducation: How climate change is taught in America.* Columbia Global Reports.

Yale Environment 360. (2019, September 4). *China's Belt and Road Initiative could drive warming to 2.7 degrees.* Yale School of the Environment. https://e360.yale.edu/digest/chinas-belt-and-road-initiative-could-drive-warming-to-2-7-degrees

Yearwood, L. (2020, June 22). *Climate justice is racial justice, racial justice is climate justice.* Shondaland. https://www.shondaland.com/act/a32905536/environmental-justice-racial-justice-marginalized-communities/

Yergin, D. (1990). *The prize: The epic quest for oil, money and power.* Simon & Schuster.

Zabel, I. H., Duggan-Haas, D. A., & Ross, R. M. (Eds). (2017). *The teacher-friendly guide to climate change.* Paleontological Research Institution.

Zakrzewski, C. (2021, November 2). Breitbart has outsize influence over climate change denial on Facebook, report says. *The Washington Post.* https://www.washingtonpost.com/technology/2021/11/02/facebook-climate-change-misinformation-breitbart/

Zerubavel, E. (1997). *Social mindscapes: An invitation to cognitive sociology.* Harvard University Press.

Zerubavel, E. (2002). The elephant in the room: Notes on the social organization of denial. In K. Cerulo (Ed.), *Culture in mind: Toward a sociology of culture and cognition* (pp. 21–27). Routledge.

Zuboff, S. (2021, January 29). The coup we are not talking about. *The New York Times.* https://www.nytimes.com/2021/01/29/opinion/sunday/facebook-surveillance-society-technology.html

Author Index

Subject Index

About the Authors

James S. Damico is a former elementary and middle school teacher from New Jersey. He currently works as a professor of Literacy, Culture, and Language Education in the Department of Curriculum & Instruction at Indiana University–Bloomington, where he is also an affiliate faculty member with the Integrated Program in the Environment and with the Center for Latin American and Caribbean Studies. His scholarship and teaching emphasize critical literacies and inquiry-based approaches for working with digital media and complex topics, especially climate change. He is the author of many journal articles and book chapters and is the coauthor of two other books, *Social Studies as New Literacies in a Global Society: Relational Cosmopolitanism in the Classroom* (2011) and *Commemorative Literacies and Labors of Justice: Resistance, Reconciliation, and Recovery in Buenos Aires and Beyond* (2022).

Mark C. Baildon is a former middle and high school social studies teacher in schools around the world (United States, Israel, Singapore, Saudi Arabia, Taiwan), a teacher educator, and a professor in Humanities and Social Studies Education in Singapore, and is currently a professor in Foundations of Education at United Arab Emirates University. His research and teaching interests focus on ways to support social studies inquiry practices, global citizenship education, and 21st-century literacies in new global contexts. He is the author of many journal articles and book chapters and has coauthored three books: *Social Studies as New Literacies in a Global Society: Relational Cosmopolitanism in the Classroom* (2011), *Controversial History Education in Asian Contexts* (2013), and *Research on Global Citizenship in Asia: Conceptions, Perceptions, and Practice* (2021).